SpringerBriefs on PDEs and Data Science

SpringerBriefs on PDEs and Data Science targets contributions that will impact the understanding of partial differential equations (PDEs), and the emerging research of the mathematical treatment of data science.

***

The series will accept high-quality original research and survey manuscripts covering a broad range of topics including analytical methods for PDEs, numerical and algorithmic developments, control, optimization, calculus of variations, optimal design, data driven modelling, and machine learning. Submissions addressing relevant contemporary applications such as industrial processes, signal and image processing, mathematical biology, materials science, and computer vision will also be considered.

The series is the continuation of a former editorial cooperation with BCAM, which resulted in the publication of 28 titles as listed here: https://www.springer.com/gp/mathematics/bcam-springerbriefs

Cristian E. Gutiérrez

Optimal Transport and Applications to Geometric Optics

Springer

Cristian E. Gutiérrez
Department of Mathematics
Temple University
Philadelphia, PA, USA

ISSN 2731-7595 ISSN 2731-7609 (electronic)
SpringerBriefs on PDEs and Data Science
ISBN 978-981-99-4866-6 ISBN 978-981-99-4867-3 (eBook)
https://doi.org/10.1007/978-981-99-4867-3

This Springer imprint is published by the registered company Springer Nature Singapore Pte Ltd.
The registered company address is: 152 Beach Road, #21-01/04 Gateway East, Singapore 189721, Singapore

Preface

This book focuses on the theory of Optimal Transport and its applications in solving problems in geometric optics. It provides a comprehensive presentation that includes a thorough analysis of key problems, namely the Monge problem, the Monge-Kantorovich problem, the transshipment problem, and the network flow problem. It also establishes the interconnections between these problems. Additionally, the book dedicates a chapter to Monge-Ampère measures, offering exercises for further understanding.

Furthermore, the book conducts a detailed analysis of the disintegration of measures and its application to the Wasserstein metric, showcasing its realization using the continuity equation. A chapter on the Sinkhorn algorithm is also included.

In terms of optics applications, the book covers the essential background knowledge on light refraction, addressing both the far-field and near-field refraction problems. It also sheds light on current research directions in this area.

The presentation of the book is self-contained, providing detailed explanations and complete proofs of the theorems and results. It is ideal for researchers, practitioners, and students interested in utilizing optimal transport principles for the design of non-rotationally symmetric lenses.

To fully grasp the content of this book, readers are expected to have a solid understanding of measure theory and integration, as well as a basic knowledge of functional analysis.

Acknowledgements

These notes has been used in graduate courses at Temple University, and I would like to thank the students Irem Altiner and Artur Andrade for reading several parts of them and for providing useful comments. I would also like to especially thank Elie Abdo for collaborating in the writing of Sect. 14. Finally, I would like to thank Juan Manfredi for his encouragement to publish these notes and to the anonymous referees for carefully reading the original manuscript and for useful comments. The author was partial supported by NSF Grant DMS-1600578.

Contents

Chapter 1
Introduction

Abstract Three problems motivating the theory of optimal transportation are introduced: the distribution problem, the Monge problem and the Kantorovich problem. Also, the network flow problem is analyzed, solved in detail and it is described how to convert it into an optimal transport problem.

We begin presenting three problems that are the motivation of the theory of optimal transportation.

1.1 The Transportation or Distribution Problem

If $X_1, \cdots, X_m$ are sources (for example warehouses) and $Y_1, \cdots, Y_n$ are destinations (for example shops), *the transportation problem* consists of transporting commodities or items from the sources to the destinations assuming the cost of transporting one item from X_i to Y_j is c_{ij}. We are assuming also that u_i is the supply at X_i and v_j is the demand at Y_j. In summary,

$$
\begin{aligned}
m &= \#\text{ of sources of goods}\\
n &= \#\text{ of destinations}\\
u_i &= \text{ capacity of source } i\\
v_j &= \text{ need or demand of destination } j\\
c_{ij} &= \text{ unit transportation cost from source } i \text{ to destination } j\\
x_{ij} &= \text{ quantity shipped from source } i \text{ to destination } j.
\end{aligned}
$$

C. E. Gutiérrez, *Optimal Transport and Applications to Geometric Optics*,
SpringerBriefs on PDEs and Data Science,
https://doi.org/10.1007/978-981-99-4867-3_1

A transportation plan is a matrix $X = (x_{ij})$ with $1 \le i \le m, 1 \le j \le n$. Each transportation plan gives rise to a cost

$$\sum_{i=1}^{m} \sum_{j=1}^{n} x_{ij}\, c_{ij} = \langle X, C\rangle.$$

The objective is then to find a transportation plan X so that the cost is minimum.

Restrictions/constraints:

1. the shipments are non negative, i.e., $x_{ij} \ge 0$;
2. capacity constraints: $\sum_{j=1}^{n} x_{ij} = u_i$ for $1 \le i \le m$;
3. needs constraints: $\sum_{i=1}^{m} x_{ij} = v_j$ for $1 \le j \le n$.

The problem is feasible if the total of sources is at least the total of needs, i.e., $\sum_{i=1}^{m} u_i \ge \sum_{i=1}^{n} v_i$. One may assume that $\sum_{i=1}^{m} u_i = \sum_{i=1}^{n} v_i$ because if $\sum_{i=1}^{m} u_i > \sum_{i=1}^{n} v_i$ we may introduce an imaginary destination Y_{n+1} with $v_{n+1} = \sum_{i=1}^{m} u_i - \sum_{i=1}^{n} v_i$ and cost $c_{i(n+1)} = 0$ for $1 \le i \le n$. That is, the excess is placed at an imaginary destination with cost zero. See [12, pp. 61–62] and [19, pp. 3–8] containing illuminating examples of application. [12, pp. 61–62] contains also a historical description and evolution of linear programming.

Let H be the Hilbert space of real matrices with m rows and n columns, $H = \mathbb{R}^{m\times n}$, with the inner product

$$\langle A, B\rangle = \text{trace } \left(A\, B^t\right).$$

Let $\mathbf{u} \in \mathbb{R}^m$ and $\mathbf{v} \in \mathbb{R}^n$ be vectors with non negative components. Consider the set of matrices $A \in \mathbb{R}^{m\times n}$ with non negative entries such that the vector sum of their rows is $\mathbf{u}$ and the vector sum of their columns is $\mathbf{v}$, that is,

$$\sum_{j=1}^{n} a_{ij} = u_i, \quad 1 \le i \le m, \text{ and } \sum_{i=1}^{m} a_{ij} = v_j, \quad 1 \le j \le n. \tag{1.1}$$

Let us denote this set of matrices by $\mathcal{N}(\mathbf{u}, \mathbf{v})$ which is referred as the transportation polytope, see [20] for many examples and the simplex method, also [35, 43]; and [8] for properties of this set of matrices; $\mathcal{N}(\mathbf{u}, \mathbf{v})$ is a compact convex set in $H = \mathbb{R}^{m\times n}$.

Therefore each matrix $A \in \mathcal{N}(\mathbf{u}, \mathbf{v})$ represents a transportation plan which yields a cost $\langle A, C\rangle$, where $C = c_{ij}$ is the cost matrix. The question is then to find a transportation plan that minimizes the total cost, that is, to find

$$\min_{A\in\mathcal{N}(\mathbf{u},\mathbf{v})} \langle A, C\rangle,$$

and a matrix A attaining this minimum. Since $\mathcal{N}(\mathbf{u}, \mathbf{v})$ is compact there must be a transportation plan that attains the minimum. However to find the optimal plan can be extremely long and computationally costly. Linear programming was invented to solve these type of problems efficiently, in particular, a method developed with this purpose is the simplex method, see [12, Chapter 5] and [19, Chapter 1].

1.2 Monge Problem

From [24]: "When we have to transport land from one place to another, we usually give the name of excavation to the volume of land that we must transport, and the name of embankment to the space they must occupy after transport. The cost of transporting a molecule being, all other things being equal, proportional to its weight and the space it is made to travel, and therefore the product of total transport must be proportional to the sum of the products of molecules multiplied by the space covered, it follows that the cut and fill being given in figure and position, it is not unimportant that such molecule of the cut is transported in such or such other place of the fill, but that 'There is a certain distribution of molecules from the first to the second, according to which the sum of these products will be the smallest possible, and the price of total transport will be a minimum."

This problem can be formally described as follows. Suppose (X, μ) and (Y, ν) are given measure spaces with $\mu(X) = \nu(Y)$, and let $c : X \times Y \to [0, +\infty)$ be a function, the cost. A function $T : X \to Y$ preserves the measures μ and ν if $\mu\left(T^{-1}(E)\right) = \nu(E)$ for each set $E \subset Y$; *T is called a transport map.* Let $\mathcal{S}(\mu, \nu)$ be the class of maps preserving μ and ν. Monge question can then be phrased as follows: Find $T \in \mathcal{S}(\mu, \nu)$ such that the integral

$$\int_X c\,(x, Tx)\; d\mu$$

is minimum among all $T \in \mathcal{S}(\mu, \nu)$. In Monge problem, the cost is the Euclidean distance $c(x, y) = |x - y|$.

At this point, all this is formal and measurability properties are needed for the precise formulation. We introduce the *push forward of the measure μ through T* by $T_{\#}\mu(E) = \mu\left(T^{-1}(E)\right)$ for $E \subset Y$. It will be proved later that $T_{\#}\mu$ is a measure and the problem above can be precisely formulated and solved under conditions on the cost c.

Notice that for certain measures μ, ν we might have $\mathcal{S}(\mu, \nu) = \emptyset$, i.e., there might not exist any measure preserving map. In fact, this is the case if for example, $X = Y = \mathbb{R}$, $\mu = \delta_0$, and $\nu = \dfrac{1}{2}\,(\delta_{-1} + \delta_1)$.

Remark 1.1 Suppose X, Y are two domains in $\mathbb{R}^n$, the measures μ and ν have continuous densities ρ and σ respectively with respect to Lebesgue measure, and $T : X \to Y$ is a measure preserving map that is a C^1 diffeomorphism. Let $\phi \in C(Y)$. From the formula of change of variables

$$\int_Y \phi(y)\,\sigma(y)\,dy = \int_X \phi(Tx)\,\sigma(Tx)\,|\det DT(x)|\,dx.$$

Since T is measure preserving $\mu\left(T^{-1}E\right) = \nu(E)$ for each Borel set $E \subset Y$ which can be rewritten as

$$\int_Y \chi_E(y)\,\sigma(y)\,dy = \int_X \chi_{T^{-1}E}(x)\,\rho(x)\,dx = \int_X \chi_E(Tx)\,\rho(x)\,dx.$$

If ϕ is a simple function, $\phi(y) = \sum_{j=1}^k \alpha_j\,\chi_{E_j}(y)$, then

$$\int_Y \phi(y)\,\sigma(y)\,dy = \int_X \phi(Tx)\,\rho(x)\,dx, \tag{1.2}$$

and since simple functions are dense in $C(Y)$ we obtain that (1.2) holds for each $\phi \in C(Y)$ (see Lemma 5.4 for a more general result). Therefore we obtain the formula

$$\int_X \phi(Tx)\,\rho(x)\,dx = \int_X \phi(Tx)\,\sigma(Tx)\,|\det DT(x)|\,dx$$

for each $\phi \in C(Y)$ which implies that T satisfies the differential equation

$$\rho(x) = \sigma(Tx)\,|\det DT(x)|.$$

1.3 Kantorovitch Problem

Let X and Y be metric spaces, (X, μ) and (Y, ν) Borel measure spaces with $\mu(X) = \nu(Y) = 1$,[1] and let $c : X \times Y \to \mathbb{R}_{\geq 0}$ be a measurable function in the product space $(X \times Y, \mu \otimes \nu)$. Consider the class $\Pi(\mu, \nu)$ of all measures γ in $X \times Y$ satisfying $\gamma(A \times Y) = \mu(A)$ for all μ-measurable subsets $A \subset X$ and $\gamma(X \times B) = \nu(B)$ for all ν-measurable subsets $B \subset Y$ (that is, the marginals of γ are μ and ν). Notice that this implies that $\gamma(X \times Y) = 1$. *The measure γ is called a transport plan.* Notice the measure $\mu \otimes \nu \in \Pi(\mu, \nu)$ so the class of admissible measures $\Pi(\mu, \nu)$ is always a non empty convex set. The Kantorovitch problem consists in minimizing

$$\int_{X \times Y} c(x, y)\,d\gamma$$

over all $\gamma \in \Pi(\mu, \nu)$.

Remark 1.2 We show that when the measures μ and ν are discrete Kantorovitch's problem is the transportation problem explained in Sect. 1.1. Indeed, let $\mu =$

[1] If $\mu(X) = \nu(Y)$ not necessarily equal one, we normalize the measures taking $\tilde{\mu} = \mu/\mu(X)$ and $\tilde{\nu} = \nu/\nu(Y)$.

$\sum_{i=1}^m u_i\,\delta_{X_i}$ and $\nu = \sum_{j=1}^n v_j\,\delta_{Y_j}$ with $\sum_{i=1}^m u_i = \sum_{j=1}^n v_j = 1$. Take $\gamma = \sum_{i=1}^m \sum_{j=1}^n u_i\, v_j\, \delta_{(X_i,Y_j)}$. Then

$$\gamma(A \times Y) = \sum_{i=1}^m \sum_{j=1}^n u_i\, v_j\, \delta_{(X_i,Y_j)}\,(A \times Y) = \sum_{i=1}^m \sum_{j=1}^n u_i\, v_j\, \delta_{X_i}\,(A)$$

$$= \sum_{j=1}^n v_j \sum_{i=1}^m u_i\, \delta_{X_i}\,(A) = \nu(Y)\,\mu(A) = \mu(A).$$

Similarly, $\gamma(X \times B) = \nu(B)$ for $B \subset Y$. So $\gamma \in \Pi(\mu,\nu)$. On the other hand, if $\pi \in \Pi(\mu,\nu)$, we shall prove that

$$\pi = \sum_{i=1}^m \sum_{j=1}^n a_{ij}\,\delta_{(X_i,Y_j)}$$

with $A = (a_{ij}) \in \mathcal{N}(\mathbf{u},\mathbf{v})$ where $\mathbf{u} = (u_1, \cdots, u_m)$ and $\mathbf{v} = (v_1, \cdots, v_n)$. Indeed, first notice that

$$\begin{aligned}\operatorname{supp}(\pi) &= \{(x, y) : \text{there is a neighborhood } N_{(x,y)} \text{ such that } \pi(N_{(x,y)}) = 0\}^c\\ &= \{(X_i, Y_j) : 1 \le i \le m, 1 \le j \le n\}\end{aligned}$$

because if $(x, y) \neq (X_i, Y_j)$ for all i, j, then $x \neq X_i$ or $y \neq Y_j$ so there is a neighborhood N_x such that $X_i \notin N_x$ and so $\pi(N_x \times Y) = \mu(N_x) = 0$, or there is a neighborhood N_y such that $Y_j \notin N_y$ and so $\pi(X \times N_y) = \nu(N_y) = 0$. Hence $\pi = \sum_{i=1}^m \sum_{j=1}^n a_{ij}\,\delta_{(X_i,Y_j)}$ with $a_{ij} = \pi\left((X_i, Y_j)\right)$.

Therefore $\Pi(\mu,\nu)$ can be identified with $\mathcal{N}(\mathbf{u},\mathbf{v})$ so

$$\int_{X\times Y} c(x, y)\,d\pi = \sum_{i=1}^m \sum_{j=1}^n a_{ij}\,c(X_i, Y_j)$$

and Kantorovitch's problem is the transportation problem.

Remark 1.3 Suppose μ and ν are probability Borel measures in X and Y respectively and let $T \in \mathcal{S}(\mu,\nu)$ be a measure preserving map, i.e., $T_{\#}\mu = \nu$. We show that T gives rise to a measure $\gamma \in \Pi(\mu,\nu)$ as follows. Let $I : X \to X$ be the identity map, and let $S : X \to X \times Y$ be defined by $Sx = (x, Tx)$. Define $\gamma_T = S_{\#}\mu$, that is, for $E \subset X \times Y$, $\gamma_T(E) = \mu\left(S^{-1}(E)\right)$. If $A \subset X$ and $B \subset Y$, then

$$\gamma_T(A \times Y) = \mu\left(S^{-1}\,(A \times Y)\right) = \mu\,(A)$$

$$\gamma_T(X \times B) = \mu\left(S^{-1}\,(X \times B)\right) = \mu\left(T^{-1}(B)\right) = \nu(B),$$

so $\gamma_T \in \Pi(\mu, \nu)$. Therefore each transport map gives rise to a transport plan. In addition,

$$\int_{X\times Y} c(x, y)\, d\gamma_T(x, y) = \int_X c(x, Tx)\, d\mu(x)$$

since $\gamma_T(SE) = \mu\left(S^{-1}(SE)\right) = \mu(E)$. Hence

$$\inf\left\{\int_{X\times Y} c(x, y)\, d\gamma : \gamma \in \Pi(\mu, \nu)\right\} \leq \inf\left\{\int_X c\,(x, Tx)\; d\mu : T \in \mathcal{S}(\mu, \nu)\right\}. \tag{1.3}$$

Existence of solutions to the Kantorovitch problem is the contents of the following theorem.

Theorem 1.4 *Let X, Y be compact metric spaces and let μ be a Borel probability measure in X and ν a Borel probability measure in Y. If the cost $c : X \times Y \to \mathbb{R}$ is continuous, then the Kantorovitch problem has a solution. That is, there exists $\gamma_0 \in \Pi(\mu, \nu)$ such that*

$$\int_{X\times Y} c(x, y)\, d\gamma_0 = \min\left\{\int_{X\times Y} c(x, y)\, d\gamma : \gamma \in \Pi(\mu, \nu)\right\}.$$

Proof We recall a few facts from functional analysis. Consider the class $C(X \times Y)$ of continuous functions over $X \times Y$ with the topology of the uniform convergence, i.e., given by the norm $\|f\|_\infty = \max_{(x,y)\in X\times Y} |f(x, y)|$. Let $C(X \times Y)^* = \{\ell : C(X \times Y) \to \mathbb{R} : \ell \text{ is linear and continuous}\}$ be the dual space with the topology given by the norm $\|\ell\|_* = \max_{\|f\|_\infty \leq 1} |\ell(f)|$. Since $X \times Y$ is compact, the dual $C(X \times Y)^*$ can be identified with the class of regular Borel measures on $X \times Y$, i.e., given $\ell \in C(X \times Y)^*$, there exists a unique regular Borel signed finite measure γ over $X \times Y$ such that $\ell(f) = \int_{X\times Y} f\, d\gamma$ for each $f \in C(X \times Y)$, see [14, Theorem 3 (Riesz representation theorem), p. 265] (if $\ell \geq 0$, then γ is a non negative measure). From Banach-Alaoglu theorem [14, Theorem 2, p. 424], the unit ball in $C(X \times Y)^*$ is compact in the weak topology. That is, if $\|\ell_j\|_* \leq 1$, there is a subsequence ℓ_{j_k} and $\ell \in C(X \times Y)^*$ such that ℓ_{j_k} converges weakly to ℓ, i.e., $\ell_{j_k}(f) \to \ell(f)$ for all $f \in C(X \times Y)$.

We introduce the following topology in $\Pi(\mu, \nu)$: $\gamma_k \to \gamma$ if $\int_{X\times Y} f(x, y)\, d\gamma_k \to \int_{X\times Y} f(x, y)\, d\gamma$ for each $f \in C(X \times Y)$. We then claim that $\Pi(\mu, \nu)$ is compact in this topology. That is, each sequence $\gamma_k \in \Pi(\mu, \nu)$ contains a subsequence γ_{k_j} and there is $\gamma \in \Pi(\mu, \nu)$ such that $\int_{X\times Y} f(x, y)\, d\gamma_{k_j} \to \int_{X\times Y} f(x, y)\, d\gamma$ for each $f \in C(X \times Y)$. Indeed, since each $\gamma \in \Pi(\mu, \nu)$ is a probability measure, to each $\gamma \in \Pi(\mu, \nu)$, there corresponds a unique non negative linear functional $\ell_\gamma \in C(X \times Y)^*$ given by $\ell_\gamma(f) = \int_{X\times Y} f\, d\gamma$. Let γ_k be a sequence in $\Pi(\mu, \nu)$. Since γ_k are all probability measures, the corresponding linear functionals ℓ_{γ_k} have norms in $C(X \times Y)^*$ all uniformly bounded. Therefore from Banach-Alaoglu theorem, ℓ_{γ_k} contains a subsequence converging weakly to a linear functional ℓ

(non negative). By Riesz' representation theorem, there exists a unique Radon measure γ_0 on $X \times Y$ with $\ell_{\gamma_0} = \ell$. In remains to show that $\gamma_0 \in \Pi(\mu, \nu)$. We have $\ell_{\gamma_k}(f) \to \ell_{\gamma_0}(f)$ for each $f \in C(X \times Y)$. If $f = f(x)$, then $\int_{X\times Y} f(x)\,d\gamma_k \to \int_{X\times Y} f(x)\,d\gamma_0$. Since $\int_{X\times Y} f(x)\,d\gamma_k = \int_{X\times Y} f(x)\,d\mu$, it follows that $\int_{X\times Y} f(x)\,d\gamma = \int_{X\times Y} f(x)\,d\mu$ for each $f \in C(X)$. Similarly, $\int_{X\times Y} g(y)\,d\gamma = \int_{X\times Y} g(y)\,d\mu$ for all $g \in C(Y)$. Since all measures appearing are regular, it follows by approximation that the marginals of γ_0 are μ and ν which means $\gamma_0 \in \Pi(\mu, \nu)$.

To complete the proof we invoke the following result: *if K is compact and $F : K \to \mathbb{R}$ is lower semicontinuous (i.e., $F(x) \leq \liminf_{k\to\infty} F(x_k)$ for $x_k \to x$), then there exists $y \in K$ such that $F(y) = \inf_{x\in K} F(x)$.* To apply this to our case, let $K = \Pi(\mu, \nu)$ with the topology described. Since $c \in C(X \times Y)$, this topology makes the function $F(\gamma) = \int_{X\times Y} c(x, y)\,d\gamma$ for $\gamma \in \Pi(\mu, \nu)$ continuous in such a topology, and the theorem follows. □

Remark 1.5 Theorem 1.4 can be extended to include costs c that are not necessarily continuous, see [50, Theorems 1.5, 1.7, and Lemma 1.8].

Remark 1.6 Kantorovitch problem can be easily extended to the multivariable case. That is, if (X_i, μ_i), $1 \leq i \leq N$, are compact Borel probability measure spaces and $c : X_1 \times \ldots \times X_N \to [0, +\infty)$ is a measurable cost function, the problem is to minimize $\int_{X_1\times\ldots\times X_N} c(x_1, \cdots, x_N)\,d\gamma$ over all measures γ in $X_1 \times \ldots \times X_N$ whose marginals are $\mu_1, \cdots, \mu_N$. Existence of solutions for this problem follows immediately by adding more variables to the argument in the proof of Theorem 1.4. See [49, Vol. 1, p. 57] and the more recently reference [2] for applications.

1.4 Trans-Shipment Problem

Let (X, d) be a compact metric space and let $\mathcal{B}$ be the Borel σ-algebra of subsets of X. Suppose $\phi : \mathcal{B} \to \mathbb{R}$ is a σ-additive set function, i.e., a finite Borel signed measure over X. By Jordan's decomposition, $\phi = \phi^+ - \phi^-$ where ϕ^+ and ϕ^- are the upper and lower variations of ϕ,[2] respectively; $\phi^\pm$ are finite Borel measures over X. Furthermore, suppose $\phi(X) = 0$, that is, $\phi^+(X) = \phi^-(X)$. Dividing by $\phi^+(X)$, we may assume $\phi^+(X) = \phi^-(X) = 1$. Let $\Gamma(\phi)$ be the class of all finite Borel measures γ over $X \times X$ such that

$$\gamma(A \times X) - \gamma(X \times A) = \phi(A),$$

[2] $\phi^+(A) = \sup_{B\subset A} \phi(B)$; $\phi^-(A) = -\inf_{B\subset A} \phi(B)$.

for all $A \in \mathcal{B}$. Clearly, the tensor product measure $\phi^+ \otimes \phi^- \in \Gamma(\phi)$.[3] The transshipment or Kantorovich-Rubinstein problem is to minimize the functional

$$\int_{X\times X} c(x,y)\,d\gamma, \quad \text{over } \gamma \in \Gamma(\phi),$$

where c is a cost function, $c \in C(X \times X)$, $c \geq 0$. Clearly, $\Pi(\phi^+,\phi^-) \subset \Gamma(\phi)$ and so

$$\inf\left\{\int_{X\times X} c(x,y)\,d\gamma : \gamma \in \Gamma(\phi)\right\} \leq \inf\left\{\int_{X\times X} c(x,y)\,d\gamma : \gamma \in \Pi(\phi^+,\phi^-)\right\}. \tag{1.4}$$

In general this inequality can be strict, but if c is a distance (and under extra assumptions) will show there is equality. If $\gamma \in \Gamma(\phi)$, then from the density of the simple functions in $C(X)$ we have that

$$\int_{X\times X} (h(x)-h(y))\,d\gamma = \int_X h(x)\,d\phi^+ - \int_X h(x)\,d\phi^-,$$

for each $h \in C(X)$. If $u, v \in C(X)$ with $u(x)+v(y) \leq c(x,y)$ for all $x, y \in X$, then

$$u(x) \leq \inf_{y\in X} (c(x,y)-v(y)) := v_c(x)$$

and $v(x) \leq -v_c(x)$ for all $x \in X$; assuming $c(x,x)=0$. We also have that $|v_c(x) - v_c(y)| \leq c(x,y)$ for all $x, y \in X$, if c is a distance. Hence

$$\begin{aligned} I(u,v) := \int_X u\,d\phi^+ + \int_X v\,d\phi^- &\leq \int_X v_c\,d\phi^+ - \int_X v_c\,d\phi^- \\ &= \int_{X\times X} (v_c(x)-v_c(y))\,d\gamma \leq \int_{X\times X} c(x,y)\,d\gamma \end{aligned}$$

and therefore

$$\begin{aligned} S &:= \sup\{I(u,v) : u, v \in C(X); u(x)+v(y) \leq c(x,y)\, \forall x, y \in X\} \\ &\leq \inf\left\{\int_{X\times X} c(x,y)\,d\gamma : \gamma \in \Gamma(\phi)\right\}. \end{aligned}$$

[3] If μ_1, μ_2 are Borel measures over X, $\mu_1 \otimes \mu_2$ is the Borel measure on $X \times X$ satisfying $(\mu_1 \otimes \mu_2)(A \times B) = \mu_1(A)\mu_2(B)$ for all A, B Borel sets of X.

Since c is a distance it is Lipschitz. Therefore we can then apply the results from Chap. 6 with $D = D^* = X$ if we assume the cost c satisfies condition (6.3) from Chap. 6 below to obtain that

$$S = \inf \left\{ \int_{X \times X} c(x, y)\, d\gamma : \gamma \in \Pi(\phi^+, \phi^-) \right\}.$$

Hence, from Remark 6.8 and (1.4) we obtain equality in (1.4). That is, the transshipment problem equals the transportation problem when the cost c is a distance satisfying (6.3).

1.5 Minimum Network Flow Problem

This is an important problem in the applications that contains as a particular case the transportation problem described in Sect. 1.1 when there are no transshipment points. It is also true, but not trivial, that the minimum network flow problem can be converted into a transportation problem. This will be shown in two different ways in Sects. 1.6 and 1.7.

Suppose we have N points (or nodes) $x_1, \cdots, x_N$ such that each point x_i is of one and only one of the following types

(a) x_i can only ship goods (*shipping point or source*)
(b) x_i can only receive goods (*receiver point or sink*)
(c) x_i can both receive and ship goods (*transshipment or switching point*).

We are also given numbers $g_1, \cdots, g_N$ that are positive, negative or zero, with

$$\sum_{i=1}^{N} g_i = 0. \tag{1.5}$$

If x_i is a shipping point, then g_i represents the *capacity* of x_i, i.e., the maximum number of goods that can be shipped from x_i; $g_i > 0$. If x_i is a recieving point, then g_i represents the *demand* at x_i, i.e., $-g_i$ is the maximum number of goods that can be received at x_i; $g_i < 0$. And if x_i is a transshipment point, then g_i is the net number of goods that are shipped/received at x_i, i.e., the difference between the goods shipped from x_i minus the goods received at x_i; so g_i can be positive, negative or zero. We assume conservation, i.e.,

total amount sent from all points=total amount received at all points.

Since the total amount sent from all points equals $\sum_{\{i:g_i>0\}} g_i$ and the total amount received at all points equals $-\sum_{\{i:g_i<0\}} g_i$, we then have

$$\sum_{\{i:g_i>0\}} g_i = -\sum_{\{i:g_i<0\}} g_i.$$

Let us introduce the following auxiliary quantities whose values are unknown at the moment:

$$a_i = \text{total amount shipped from } x_i \text{ to all other points}$$

and

$$b_i = \text{total amount received at } x_i \text{ from all other points,}$$

so we have

$$g_i = a_i - b_i,$$

and obviously, $a_i, b_i \geq 0$. Notice that if x_i is a shipping point, then $b_i = 0$; and if x_i is a receiving point, then $a_i = 0$. Let

$$c_{ij}\text{=cost of shipping one unit of goods from } x_i \text{ to } x_j, i \neq j, \tag{1.6}$$

with $c_{ii} = 0$. Also

$$c_{ji}\text{=cost of receiving one unit of goods at } x_i \text{ from } x_j, i \neq j.$$

Let

$$x_{ij} = \text{\# of units shipped from } x_i \text{ to } x_j.$$

The total number of units shipped from x_i to all points is then

$$\sum_{j=1}^{N} x_{ij} = a_i.$$

On the other hand, the total number of units received at x_i from all points is

$$\sum_{j=1}^{N} x_{ji} = b_i.$$

Therefore, the total cost of shipping/receiving to/from all points is then

$$\sum_{i,j=1}^{N} x_{ij}\, c_{ij},$$

and so *the minimum network flow problem* is to minimize this quantity over all $x_{ij} \geq 0$ satisfying the condition

$$\sum_{j=1}^{N} x_{ij} - \sum_{j=1}^{N} x_{ji} = g_i, \qquad 1 \leq i \leq N. \tag{1.7}$$

Example Suppose we have four points x_1, x_2, x_3, x_4 so that x_1, x_2 are shipping points, x_3 is a receiver point, and x_4 is a transshipment point. Also x_{ij} = # of units shipped from x_i to x_j. We have

(at x_1) we have $x_{11} = 0$, $x_{12} = 0$ since x_2 is shipping point. The amount shipped from x_1 equals $x_{13}+x_{14}$. Since x_1 is a shipping point, $x_{21} = x_{31} = x_{41} = 0$. So the net flow at x_1 equals $x_{13} + x_{14} = g_1$.

(at x_2) we have $x_{22} = 0$, $x_{21} = 0$ since x_1 is shipping point. The amount shipped from x_2 equals $x_{23} + x_{24}$. Since x_2 is a shipping point, $x_{12} = 0$. Since x_3 is receiving $x_{32} = 0$, and since x_2 is shipping, $x_{42} = 0$. So the net flow at x_2 equals $x_{23} + x_{24} = g_2$.

(at x_3) Since x_3 is receiving only, we have $x_{31} = x_{32} = x_{34} = 0$ and so the amount shipped from x_3 is zero. The amount received at x_3 equals $x_{13} + x_{23} + x_{43}$. So the net flow at x_3 equals $0 - (x_{13} + x_{23} + x_{43}) = g_3$.

(at x_4) Since x_4 is a transshipment point, we have the amount shipped from x_4 equals x_{43}, since $x_{41} = x_{42} = 0$, because x_1, x_2 are only shipping points. The amount received at x_4 equals $x_{14} + x_{24}$. So the net flow at x_4 equals $x_{43} - (x_{14} + x_{24}) = g_4$.

Summarizing, we have the equations

$$\begin{aligned} x_{13} + x_{14} &= g_1 \\ x_{23} + x_{24} &= g_2 \\ -(x_{13} + x_{23} + x_{43}) &= g_3 \\ x_{43} - (x_{14} + x_{24}) &= g_4. \end{aligned}$$

That is, there are five unknowns $x_{13}, x_{14}, x_{23}, x_{24}, x_{43}$, all nonnegative, that must be chosen to maximize the quantity

$$c_{13}\, x_{13} + c_{14}\, x_{14} + c_{23}\, x_{23} + c_{24}\, x_{24} + c_{43}\, x_{43}.$$

In general, suppose we have N points such that

$$\begin{aligned} \mathcal{S} &= \{x_1, \cdots, x_n\} \quad \text{are } n \text{ shipping points} \\ \mathcal{R} &= \{x_{n+1}, \cdots, x_{n+m}\} \quad \text{are } m \text{ receiving points} \\ \mathcal{T} &= \{x_{n+m+1}, \cdots, x_N\} \quad \text{are } N-(n+m) \text{ transshipment points.} \end{aligned}$$

If from the point x_i one can ship to x_j, we denote this by the pair (x_i, x_j), and say it is an admissible pair. So all pairs in $\mathcal{S} \times (\mathcal{R} \cup \mathcal{T})$ are admissible, as well as the pairs in the set $\mathcal{T} \times \mathcal{R}$, and also in $(\mathcal{T} \times \mathcal{T}) \setminus D(\mathcal{T})$ where $D(\mathcal{T}) = \{(x_i, x_i) : x_i \in \mathcal{T}\}$. Therefore, the set of all admissible pairs is the set

$$\mathcal{A} = (\mathcal{S} \times (\mathcal{R} \cup \mathcal{T})) \cup (\mathcal{T} \times \mathcal{R}) \cup ((\mathcal{T} \times \mathcal{T}) \setminus D(\mathcal{T})) .$$

In the previous example, $\mathcal{S}$ has two points, and $\mathcal{R}$ and $\mathcal{T}$ have both one point, and so $\mathcal{A}$ has five points in agreement with the aforementioned example. In general,

$$\begin{aligned} \#(\mathcal{S} \times (\mathcal{R} \cup \mathcal{T})) &= n\,(N-n) \\ \#(\mathcal{T} \times \mathcal{R}) &= (N-(n+m))\,m \\ \#((\mathcal{T} \times \mathcal{T}) \setminus D(\mathcal{T})) &= (N-(n+m))^2 - (N-(n+m)), \end{aligned}$$

and since the three sets are disjoint

$$\begin{aligned} \#(\mathcal{A}) &= n\,(N-n) + (N-(n+m))\,m + (N-(n+m))^2 - (N-(n+m)) \\ &= n\,(N-n) + (N-n-m)\,(N-n-1) . \end{aligned}$$

In addition,

$$x_{ij} = 0 \qquad \text{for } n+1 \le i \le n+m \text{ and } 1 \le j \le N$$

since x_i is a sink for $n+1 \le i \le n+m$; and

$$x_{ij} = 0 \qquad \text{for } n+1 \le i \le N \text{ and } 1 \le j \le n \text{ since } x_j \text{ is only shipping;}$$

and also $x_{ii} = 0$ for $1 \le i \le N$.

1.6 Conversion of the Network Flow Problem Into a Transportation Problem

At a shipping point x_i we have $g_i > 0$, $a_i = g_i$ and $b_i = 0$. Also, at a receiving point x_i we have $g_i < 0$, $a_i = 0$ and $b_i = -g_i$. Let us analyze what happens at a transshipment point x_i. Since $g_i = a_i - b_i$ and $a_i, b_i \geq 0$, we obviously have

$$-b_i \leq g_i \leq a_i. \tag{1.8}$$

Since we only have that the difference $a_i - b_i$ is prescribed, the values a_i and b_i could be arbitrarily large. Suppose that $g_i > 0$. Then from (1.8), the smallest value that a_i can take is g_i, and the smallest value b_i can take is 0. On the other hand, if $g_i < 0$, then again from (1.8), the smallest value that a_i can take is 0, and the smallest value b_i can take is $-g_i$.

Let $\tau = \#(\mathcal{T})$ and let $s_1, \cdots, s_\tau$ be nonnegative numbers, meaning buffer storages or cushions at each transshipment point. To each transshipment point x_i we assign the following capacity and demand:

$$\begin{aligned} &\text{if } g_i > 0, \text{ then the capacity of } x_i \text{ equals } u_i = g_i \\ &\qquad + s_i \text{ and the demand at } x_i \text{ equals } v_i = s_i \\ &\text{if } g_i \leq 0, \text{ then the capacity of } x_i \text{ equals } u_i \\ &\qquad = s_i \text{ and the demand at } x_i \text{ equals } v_i = -g_i + s_i. \end{aligned} \tag{1.9}$$

We then consider the transportation problem from Sect. 1.1 between the sets $X = \mathcal{S} \cup \mathcal{T}$ and $Y = \mathcal{R} \cup \mathcal{T}$; $\#(X) = N - n$, $\#(Y) = N - m$. Notice that $\mathcal{T}$ plays the role of sources and destinations. Denote by u_i the capacity of the source $x_i \in X$ defined as follows: if $x_i \in \mathcal{S}$, then $u_i = g_i$ and if $x_i \in \mathcal{T}$ then u_i is defined as before depending on the sign of g_i; so the vector (u_i) has $N - n$ components. The demand v_i at the destination is $v_i = -g_i$ when $x_i \in \mathcal{R}$ and when $x_i \in \mathcal{T}$ is defined as before depending on the sign of g_i; the vector (v_i) has $N - m$ components. Clearly these capacities and demands depend on the choice of $s_1, \cdots, s_\tau$. The cost function for this problem is c_{ij} = cost of transporting one unit from $x_i \in \mathcal{S} \cup \mathcal{T}$ to $x_j \in \mathcal{R} \cup \mathcal{T}$ as defined by (1.6), i.e., the same cost as for the network flow problem. With these definitions of capacities and demands, we show from (1.5) that the balance condition $\sum u_i = \sum v_i$ holds. Indeed,

$$\begin{aligned} \sum u_i &= \sum_{\{i:x_i \in \mathcal{S}\}} u_i + \sum_{\{i:x_i \in \mathcal{T}\}} u_i \\ &= \sum_{\{i:x_i \in \mathcal{S}\}} g_i + \sum_{\{i:x_i \in \mathcal{T}, g_i \geq 0\}} u_i + \sum_{\{i:x_i \in \mathcal{T}, g_i < 0\}} u_i \end{aligned}$$

$$= \sum_{\{i:x_i \in \mathcal{S}\}} g_i + \sum_{\{i:x_i \in \mathcal{T}, g_i \geq 0\}} (g_i + s_i) + \sum_{\{i:x_i \in \mathcal{T}, g_i < 0\}} s_i$$

$$= \sum_{\{i:x_i \in \mathcal{S}\}} g_i + \sum_{\{i:x_i \in \mathcal{T}, g_i \geq 0\}} g_i + \sum_{\{i:x_i \in \mathcal{T}\}} s_i,$$

and

$$\sum v_i = \sum_{\{i:x_i \in \mathcal{R}\}} v_i + \sum_{\{i:x_i \in \mathcal{T}\}} v_i$$

$$= \sum_{\{i:x_i \in \mathcal{R}\}} -g_i + \sum_{\{i:x_i \in \mathcal{T}, g_i \geq 0\}} v_i + \sum_{\{i:x_i \in \mathcal{T}, g_i < 0\}} v_i$$

$$= \sum_{\{i:x_i \in \mathcal{R}\}} -g_i + \sum_{\{i:x_i \in \mathcal{T}, g_i \geq 0\}} s_i + \sum_{\{i:x_i \in \mathcal{T}, g_i < 0\}} (-g_i + s_i)$$

$$= \sum_{\{i:x_i \in \mathcal{R}\}} -g_i + \sum_{\{i:x_i \in \mathcal{T}, g_i < 0\}} -g_i + \sum_{\{i:x_i \in \mathcal{T}\}} s_i.$$

From (1.5)

$$\sum_{\{i:x_i \in \mathcal{S}\}} g_i + \sum_{\{i:x_i \in \mathcal{T}, g_i \geq 0\}} g_i = \sum_{\{i:x_i \in \mathcal{R}\}} -g_i + \sum_{\{i:x_i \in \mathcal{T}, g_i < 0\}} -g_i$$

and so the desired identity follows.

Therefore, from Sect. 1.1, there exists a solution to the transportation problem between X and Y, with capacity and demands u_i and v_i, respectively. Notice that this problem and its solution depend on the choice of $s_1, \cdots, s_\tau$. Let us denote this transportation problem by $T(\mathbf{s})$, $\mathbf{s} = (s_1, \cdots, s_\tau)$. We denote its solution by x'_{ij} that minimizes $\sum_{ij} x_{ij} c_{ij}$ over all $x_{ij} \geq 0$ such that

$$\sum_{\{j:x_j \in Y\}} x_{ij} = u_i \text{ for all } i \text{ with } x_i \in X,$$

and

$$\sum_{\{i:x_i \in X\}} x_{ij} = v_j \text{ for all } j \text{ with } x_j \in Y.$$

The matrix $x' = (x'_{ij})$ has $N - m$ rows and $N - n$ columns, and represented in blocks we write it as

$$x' = \begin{bmatrix} \mathcal{S} \times \mathcal{R} \; \mathcal{S} \times \mathcal{T} \\ \mathcal{T} \times \mathcal{R} \; \mathcal{T} \times \mathcal{T} \end{bmatrix},$$

and we extend it to an $N \times N$ matrix $\bar{x} = (\bar{x}_{ij})$ as follows

$$\bar{x}_{ij} = \begin{cases} x'_{ij} & \text{if } (x_i, x_j) \in (\mathcal{S} \times (\mathcal{R} \cup \mathcal{T})) \cup (\mathcal{T} \times (\mathcal{R} \cup \mathcal{T})) \\ 0 & \text{otherwise,} \end{cases}$$

i.e.,

$$\bar{x} = \begin{bmatrix} 0 \; \mathcal{S} \times \mathcal{R} \; \mathcal{S} \times \mathcal{T} \\ 0 \quad 0 \qquad 0 \\ 0 \; \mathcal{T} \times \mathcal{R} \; \mathcal{T} \times \mathcal{T} \end{bmatrix}.$$

Recall that if x_{ij} represents the number of units shipped from $x_i \in X \cup Y$ to $x_j \in X \cup Y$, then $x_{ij} = 0$ if $x_i \in \mathcal{S} \cup \mathcal{R} \cup \mathcal{T}$, and $x_j \in \mathcal{S}$, $i \neq j$, since points in $\mathcal{S}$ are only shipping; and $x_{ij} = 0$ if $x_i \in \mathcal{R}$, $x_j \in \mathcal{S} \cup \mathcal{R} \cup \mathcal{T}$, since points in $\mathcal{R}$ are only sinks.

With $\bar{x}_{ij}$ define

$$y_{ij} = \begin{cases} \bar{x}_{ij} & \text{if } i \neq j \\ 0 & \text{if } i = j. \end{cases}$$

We relate y_{ij} to the network flow problem. We first verify that y_{ij} satisfies (1.7). Since x'_{ij} solves the transport problem $T(\mathbf{s})$

$$\sum_{\{j: x_j \in \mathcal{R} \cup \mathcal{T}\}} x'_{ij} = u_i \qquad \text{for each } i \text{ such that } x_i \in \mathcal{S} \cup \mathcal{T} \tag{1.10}$$

and

$$\sum_{\{i: x_i \in \mathcal{S} \cup \mathcal{T}\}} x'_{ij} = v_j \qquad \text{for each } j \text{ such that } x_j \in \mathcal{R} \cup \mathcal{T}. \tag{1.11}$$

To verify (1.7) we proceed by cases.

Case 1 $x_i \in \mathcal{S}$. Write

$$\begin{aligned} \sum_{j=1}^{N} y_{ij} - \sum_{j=1}^{N} y_{ji} &= \sum_{\{j: x_j \in \mathcal{S} \cup \mathcal{R} \cup \mathcal{T}\}} \left(y_{ij} - y_{ji} \right) \\ &= \sum_{\{j: x_j \in \mathcal{S}\}} y_{ij} + \sum_{\{j: x_j \in \mathcal{R}\}} y_{ij} + \sum_{\{j: x_j \in \mathcal{T}\}} y_{ij} - \sum_{\{j: x_j \in \mathcal{S}\}} y_{ji} \\ &\quad - \sum_{\{j: x_j \in \mathcal{R}\}} y_{ji} - \sum_{\{j: x_j \in \mathcal{T}\}} y_{ji} \end{aligned}$$

$$= \sum_{\{j:x_j \in \mathcal{S}, j \neq i\}} x'_{ij} + \sum_{\{j:x_j \in \mathcal{R}\}} x'_{ij} + \sum_{\{j:x_j \in \mathcal{T}\}} x'_{ij} - \sum_{\{j:x_j \in \mathcal{S}, j \neq i\}} x'_{ji}$$

$$- \sum_{\{j:x_j \in \mathcal{R}\}} x'_{ji} - \sum_{\{j:x_j \in \mathcal{T}\}} x'_{ji}$$

$$= 0 + u_i - 0 - 0 - 0 \quad \text{from (1.10) and since } \mathcal{R} \text{ is sink and } \mathcal{S} \text{ is shipping}$$

$$= g_i.$$

Case 2 $x_i \in \mathcal{R}$. Write

$$\sum_{j=1}^{N} y_{ij} - \sum_{j=1}^{N} y_{ji} = \sum_{\{j:x_j \in \mathcal{S} \cup \mathcal{R} \cup \mathcal{T}\}} (y_{ij} - y_{ji})$$

$$= \sum_{\{j:x_j \in \mathcal{S}\}} y_{ij} + \sum_{\{j:x_j \in \mathcal{R}\}} y_{ij} + \sum_{\{j:x_j \in \mathcal{T}\}} y'_{ij} - \sum_{\{j:x_j \in \mathcal{S}\}} y_{ji}$$

$$- \sum_{\{j:x_j \in \mathcal{R}\}} y_{ji} - \sum_{\{j:x_j \in \mathcal{T}\}} y_{ji}$$

$$= \sum_{\{j:x_j \in \mathcal{S}, j \neq i\}} \bar{x}_{ij} + \sum_{\{j:x_j \in \mathcal{R}, j \neq i\}} \bar{x}_{ij} + \sum_{\{j:x_j \in \mathcal{T}\}} \bar{x}_{ij} - \sum_{\{j:x_j \in \mathcal{S}\}} \bar{x}_{ji}$$

$$- \sum_{\{j:x_j \in \mathcal{R}, j \neq i\}} \bar{x}_{ji} - \sum_{\{j:x_j \in \mathcal{T}\}} \bar{x}_{ji}$$

$$= 0 + 0 + 0 - \sum_{\{j:x_j \in \mathcal{S}\}} \bar{x}_{ji} - 0 - \sum_{\{j:x_j \in \mathcal{T}\}} \bar{x}_{ji} \quad \text{since } \mathcal{R} \text{ is sink}$$

$$= - \sum_{\{j:x_j \in \mathcal{S} \cup \mathcal{T}\}} x'_{ji}$$

$$= -v_i = -(-g_i) = g_i \quad \text{from (1.11) and the definition of } v_i.$$

Case 3 $x_i \in \mathcal{T}$. Write

$$\sum_{j=1}^{N} y_{ij} - \sum_{j=1}^{N} y_{ji} = \sum_{\{j:x_j \in \mathcal{S} \cup \mathcal{R} \cup \mathcal{T}\}} (y_{ij} - y_{ji})$$

$$= \sum_{\{j:x_j \in \mathcal{S}\}} y_{ij} + \sum_{\{j:x_j \in \mathcal{R}\}} y_{ij} + \sum_{\{j:x_j \in \mathcal{T}\}} y'_{ij} - \sum_{\{j:x_j \in \mathcal{S}\}} y_{ji} - \sum_{\{j:x_j \in \mathcal{R}\}} y_{ji} - \sum_{\{j:x_j \in \mathcal{T}\}} y_{ji}$$

$$= \sum_{\{j:x_j\in\mathcal{S},j\neq i\}} \bar{x}_{ij} + \sum_{\{j:x_j\in\mathcal{R}\}} \bar{x}_{ij} + \sum_{\{j:x_j\in\mathcal{T},j\neq i\}} \bar{x}_{ij} - \sum_{\{j:x_j\in\mathcal{S}\}} \bar{x}_{ji} - \sum_{\{j:x_j\in\mathcal{R},j\neq i\}} \bar{x}_{ji}$$

$$- \sum_{\{j:x_j\in\mathcal{T},j\neq i\}} \bar{x}_{ji}$$

$$= 0 + \sum_{\{j:x_j\in\mathcal{R}\}} \bar{x}_{ij} + \sum_{\{j:x_j\in\mathcal{T},j\neq i\}} \bar{x}_{ij} - \sum_{\{j:x_j\in\mathcal{S}\}} \bar{x}_{ji}$$

$$- 0 - \sum_{\{j:x_j\in\mathcal{T},j\neq i\}} \bar{x}_{ji} \quad \text{since } \mathcal{S} \text{ is shipping and } \mathcal{R} \text{ is sink}$$

$$= \sum_{\{j:x_j\in\mathcal{R}\cup\mathcal{T} j\neq i\}} \bar{x}_{ij} - \sum_{\{j:x_j\in\mathcal{S}\cup\mathcal{T},j\neq i\}} \bar{x}_{ji} = \sum_{\{j:x_j\in\mathcal{R}\cup\mathcal{T}\}} x'_{ij} - \sum_{\{j:x_j\in\mathcal{S}\cup\mathcal{T}\}} x'_{ji}$$

$$= u_i - v_i \quad \text{from (1.10) and (1.11)}$$

$$= \begin{cases} g_i + s_i - s_i & \text{if } g_i > 0 \\ s_i - (-g_i + s_i) & \text{if } g_i \leq 0 \end{cases} \quad \text{from the definitions of } u_i \text{ and } v_i \text{ for } x_i \in \mathcal{T}$$

$$= g_i.$$

This concludes the verification that y_{ij} satisfies (1.7) meaning that y_{ij} is among the admissible plans for the network flow problem and therefore

$$\text{minimum cost of the network flow problem} \leq \sum_{ij} y_{ij}\, c_{ij} = \text{minimum of } T(\mathbf{s}) \tag{1.12}$$

for any $\mathbf{s}$ with nonnegative components.

Let $\mathbf{s} = (s_1, \cdots, s_\tau)$ with $s_i \geq 0$, and let

$$\mathbf{u(s)} = (u_1, \cdots, u_n, u_{n+1}, \cdots, u_{N-n})$$

and

$$\mathbf{v(s)} = (v_1, \cdots, v_n, v_{n+1}, \cdots, v_{N-m})$$

be defined by (1.9). Let $\mathbf{s}, \mathbf{t}$ be two vectors with components $s_i \leq t_i$ for $1 \leq i \leq \tau$. We shall prove that

$$m(\mathbf{t}) := \text{minimum of } T(\mathbf{t}) \leq m(\mathbf{s}) := \text{minimum of } T(\mathbf{s}). \tag{1.13}$$

First notice that $u_i(\mathbf{s}) \le u_i(\mathbf{t})$ for $1 \le i \le N - m$, and $v_j(\mathbf{s}) \le v_j(\mathbf{t})$ for $1 \le j \le N - n$. Because if $x_i \in \mathcal{S}$, then $u_i(\mathbf{s}) = u_i(\mathbf{t}) = g_i$, and $x_i \in \mathcal{T}$ and $g_i > 0$, then $u_i(\mathbf{t}) = g_i + t_i = g_i + t_i + s_i - s_i = u_i(\mathbf{s}) + t_i - s_i$, and $v_i(\mathbf{t}) = t_i = t_i - s_i + s_i = v_i(\mathbf{s}) + t_i - s_i$. And similarly when $g_i < 0$. Let x_{ij} be the optimal plan such that $m(\mathbf{s}) = \sum_{ij} x_{ij}\, c_{ij}$. We have $\sum_j x_{ij} = u_i(\mathbf{s})$ for $1 \le i \le N - m$ and $\sum_i x_{ij} = v_j(\mathbf{s})$ for $1 \le j \le N - n$. Let

$$\bar{x}_{ij} = \begin{cases} x_{ij} & \text{if } (x_i, x_j) \in [(\mathcal{S} \times (\mathcal{R} \times \mathcal{T})) \cup (\mathcal{T} \times (\mathcal{R} \times \mathcal{T}))] \setminus D(\mathcal{T}) \\ x_{ii} + t_i - s_i & \text{if } x_i \in \mathcal{T}. \end{cases}$$

We claim that $\bar{x}_{ij}$ is an admissible plan for $T(\mathbf{t})$. Indeed, if $x_i \in \mathcal{T}$,

$$\begin{aligned} \sum_j \bar{x}_{ij} &= \sum_{j \neq i} \bar{x}_{ij} + \bar{x}_{ii} = \sum_{j \neq i} x_{ij} + x_{ii} + t_i - s_i \\ &= \sum_j x_{ij} + t_i - s_i = u_i(\mathbf{s}) + t_i - s_i = u_i(\mathbf{t}). \end{aligned}$$

If $x_i \in \mathcal{S}$,

$$\sum_j \bar{x}_{ij} = \sum_j x_{ij} = u_i(\mathbf{s}) = u_i(\mathbf{t}).$$

Similarly, $\sum_i \bar{x}_{ij} = v_j(\mathbf{t})$. Since $c_{ii} = 0$, we have

$$\text{minimum of } T(\mathbf{t}) \le \sum_{ij} \bar{x}_{ij}\, c_{ij} = \sum_{ij} x_{ij}\, c_{ij} = \text{minimum of } T(\mathbf{s})$$

obtaining (1.13).

Let us consider the set

$$E = \{s_0 \in [0, +\infty) : \text{there exists an optimal plan } x_{ij} \text{ for } T(\mathbf{s}_0) \text{ such that } x_{ii} \neq 0 \text{ for } 1 \le i \le \tau\}.$$

Notice that if $0 \le s < \infty$ and $s \notin E$, then any optimal plan for $T(s)$ satisfies $x_{kk} = 0$ for some k with $x_k \in \mathcal{T}$. Then

$$\begin{aligned} \text{minimum of } T(s) = \sum_{i,j} x_{ij}\, c_{ij} &= \sum_{\{j : x_j \in \mathcal{R} \cup \mathcal{T}\}} x_{kj}\, c_{kj} \ge \sum_{\{j : x_j \in \mathcal{R} \cup \mathcal{T}, j \neq k\}} x_{kj}\, c_{kj} \\ &\ge c_0 \sum_{\{j : x_j \in \mathcal{R} \cup \mathcal{T}, j \neq k\}} x_{kj} = c_0 \sum_{\{j : x_j \in \mathcal{R} \cup \mathcal{T}\}} x_{kj} \\ &= c_0\, u_k \ge c_0\, s. \end{aligned}$$

Since c_{ij} are all strictly positive for $k \neq j$, the constant $c_0 > 0$ is such that $c_{kj} \geq c_0$ for all $k \neq j$. If E were empty, then from (1.13) this implies a contradiction. To show E is unbounded suppose E would be bounded. So $E \subset [0, M]$ for some $M \geq 0$. This means that given $s > M$, each optimal plan for $T(\mathbf{s})$ has at least one k with $x_{kk} = 0$. From the last argument

$$m(s) = \text{minimum of } T(s) \geq c_0\, s$$

but from (1.13) we obtain $m(M) \geq c_0\, s$ for each $s > M$, a contradiction.

We prove now that there exists $0 \leq s_0 < \infty$ such that a corresponding transport plan x_{ij} giving the minimum of the problem $T(\mathbf{s}_0)$, with $\mathbf{s}_0$ the vector having all coordinates equal s_0, satisfies $x_{ii} \neq 0$ for all i with $x_i \in \mathcal{T}$.

For any vector $\mathbf{s}$ with nonnegative components, if x_{ij} is an admissible transport plan for $T(\mathbf{s})$, then $s_i \geq x_{ii} \geq 0$ for all i such that $x_i \in \mathcal{T}$. In fact, since x_{ij} is admissible, if $x_i \in \mathcal{T}$ and $g_i > 0$, then from (1.9) $v_i(\mathbf{s}) = s_i$, and from (1.11) $x_{ii} \leq \sum_j x_{ji} = v_i(\mathbf{s}) = s_i$. On the other hand, if $x_i \in \mathcal{T}$ and $g_i \leq 0$, then from (1.9) $u_i(\mathbf{s}) = s_i$, and from (1.10) $x_{ii} \leq \sum_j x_{ij} = u_i(\mathbf{s}) = s_i$.

Fix a vector $\mathbf{s}$ with nonnegative components, and let x'_{ij} be the optimal transport plan for $T(\mathbf{s})$. Let $\mathbf{s}' = \left(s_1 - x'_{11}, s_2 - x'_{22}, \cdots, s_\tau - x'_{\tau\tau}\right)$. We claim that

$$\text{minimum of } T(\mathbf{s}') = \text{minimum of } T(\mathbf{s}). \tag{1.14}$$

In fact, let

$$x''_{ij} = \begin{cases} x'_{ij} & \text{for } i \neq j \\ 0 & \text{for } i = j. \end{cases}$$

We have that x''_{ij} is an admissible plan for $T(\mathbf{s}')$ because $\sum_j x''_{ij} = u_i(\mathbf{s}) - x'_{ii} = u_i(\mathbf{s}')$ and $\sum_j x''_{ji} = v_i(\mathbf{s}) - x'_{ii} = v_i(\mathbf{s}')$, for i with $x_i \in \mathcal{T}$. Also, since $c_{ii} = 0$,

$$\text{minimum of } T(\mathbf{s}) = \sum_{ij} x'_{ij}\, c_{ij} = \sum_{ij} x''_{ij}\, c_{ij} \geq \text{minimum of } T(\mathbf{s}').$$

Viceversa, let x'_{ij} be an optimal transport plan for $T(\mathbf{s})$ and let $\bar{x}_{ij}$ be an optimal transport plan for $T(\mathbf{s}')$, then $\sum_j \bar{x}_{ij} = u_i(\mathbf{s}')$ and $\sum_j \bar{x}_{ji} = v_i(\mathbf{s}')$. Define

$$x^*_{ij} = \begin{cases} \bar{x}_{ij} & \text{for } i \neq j \\ \bar{x}_{ii} + x'_{ii} & \text{for } i = j. \end{cases}$$

Then from the definition (1.9), x^*_{ij} is an admissible plan for $T(\mathbf{s})$, and since $c_{ii} = 0$,

$$\text{minimum of } T(\mathbf{s}) \le \sum_{ij} x^*_{ij}\, c_{ij} = \sum_{ij} \bar{x}_{ij}\, c_{ij} = \text{minimum of } T(\mathbf{s}').$$

This completes the proof of (1.14).

Suppose y_{ij} is the optimal solution for the minimum network flow problem. For $1 \le i \le N$, we have

$$u_i := \sum_{\{j: x_j \in \mathcal{R} \cup \mathcal{T}\}} y_{ij} = \begin{cases} g_i & \text{for } i \text{ with } x_i \in \mathcal{S} \\ s_i & \text{for } i \text{ with } x_i \in \mathcal{T}; \\ 0 & \text{for } i \text{ with } x_i \in \mathcal{R} \end{cases}$$

$$v_i := \sum_{\{j: x_j \in \mathcal{S} \cup \mathcal{T}\}} y_{ji} = \begin{cases} 0 & \text{for } i \text{ with } x_i \in \mathcal{S} \\ r_i & \text{for } i \text{ with } x_i \in \mathcal{T} \\ -g_i & \text{for } i \text{ with } x_i \in \mathcal{R} \end{cases}$$

Since y_{ij} is an admissible plan for the network flow problem, it satisfies (1.7). Suppose also the $g_i = 0$ for $x_i \in \mathcal{T}$, and hence $s_i = r_i$. Also $u_i - v_i = g_i$ for $x_i \in \mathcal{S}$ and $u_i - v_i = -g_i$ for $x_i \in \mathcal{R}$. From (1.5) $\sum_i u_i = \sum_i v_i$. If we then consider the transport problem $T(s_1, \cdots, s_\tau)$ with capacity and demands vectors u_i and v_i respectively, this means that y_{ij} is an admissible plan for this problem and so

$$\sum_{ij} y_{ij}\, c_{ij} \ge \text{minimum of } T(s_1, \cdots, s_\tau).$$

Now take any $\bar{s}_i \ge s_i$ for $1 \le i \le \tau$, then from (1.13) we have that

$$\text{minimum of } T(s_1, \cdots, s_\tau) \ge \text{minimum of } T(\bar{s}_1, \cdots, \bar{s}_\tau),$$

and from (1.12) we obtain

$$\sum_{ij} y_{ij}\, c_{ij} = \text{minimum of } T(\bar{s}_1, \cdots, \bar{s}_\tau),$$

for any $\bar{\mathbf{s}} \ge \mathbf{s}$. Therefore the minimum of the network flow problem is the minimum of a transportation problem.

Let $s_0 \in E$. So there is an optimal plan x_{ij} for $T(\mathbf{s}_0)$, with $x_{ii} \ne 0$ for all i with $x_i \in \mathcal{T}$. Consider the box $B = \prod_{i=1}^{\tau}[s_0 - x_{ii}, s_0]$, we have that $m(\mathbf{s}) = m(\mathbf{s}_0)$ for each $\mathbf{s} \in B$. Because if $\mathbf{s}' \le \mathbf{s} \le \mathbf{s}_0$ with $\mathbf{s}' = \mathbf{s}_0 - (x_{11}, \cdots, x_{\tau\tau})$, then $m(\mathbf{s}_0) \le m(\mathbf{s}) \le m(\mathbf{s}') = m(\mathbf{s}_0)$ by (1.13) and (1.14). That is, $m(\mathbf{s})$ is constant in the box B. Since E is unbounded, we can pick $\bar{s} \in E$ arbitrarily large, in particular, we can pick $\bar{s} \ge s_i$ for $1 \le i \le \tau$, and therefore the minimum of network flow problem equals the minimum of any transport problem with parameter $\bar{s}$ sufficiently large.

1.7 Another Way to Convert the Network Flow Problem Into Optimal Transport

Suppose $\mathcal{S} = \{X_1, \cdots, X_n\}$ are sources, $\mathcal{R} = \{Y_1, \cdots, Y_m\}$ are sinks, and $\mathcal{T} = \{T_1, \cdots, T_k\}$ are transshipment points. From points in $\mathcal{S}$ one can ship goods to any other points except to points in $\mathcal{S}$, and at points in $\mathcal{R}$ one can only receive. From points in $\mathcal{T}$ one can ship to $\mathcal{R}$, and receive from $\mathcal{S}$, and from same or all points in $\mathcal{T}$ one can ship and/or receive within themselves.[4] Given $X, Y \in \mathcal{N} := \mathcal{S} \cup \mathcal{T} \cup \mathcal{R}$, we say X *is connected to* Y if one can ship goods from X to Y and let

$$\mathcal{A} = \{(X, Y) \in \mathcal{N} \times \mathcal{N} : X \text{ is connected to } Y\}.$$

So with our configuration we have

$$\mathcal{S} \times (\mathcal{T} \cup \mathcal{R}) \subset \mathcal{A}, \quad \mathcal{T} \times \mathcal{R} \subset \mathcal{A},$$

and the set $\mathcal{T} \times \mathcal{T}$ may be partially or fully contained in $\mathcal{A}$. With this set up and a cost $c(X, Y)$ consider the minimum network flow problem with capacities and demands g_X for $X \in \mathcal{S} \cup \mathcal{R}$, $\sum_X g_X = 0$, $g_X > 0$ for $X \in \mathcal{S}$, $g_X < 0$ for $X \in \mathcal{R}$; $g_X = 0$ for $X \in \mathcal{T}$. We will show that its minimum equals the minimum of a transportation problem between $\mathcal{A}$, considered as sources, and $\mathcal{N}$, considered as destinations, with the following capacity or supply at each source, and demand at each destination; and with a cost to be defined in a moment. For each $(X, Y) \in \mathcal{A}$ we introduce a nonnegative number $\alpha(X, Y)$ denoting the maximum number of goods that can be sent from X to Y, called the capacity of the arc (X, Y). Let us assume these capacities are sufficiently large (in comparison with all g_X); they can be taken, for example, all equal to a large positive constant. For each arc $(X, Y) \in \mathcal{A}$ we then assign its capacity as $\alpha(X, Y)$ and for each $X \in \mathcal{N}$ the following demand

$$v_X = \begin{cases} \sum_{\{Z:(X,Z)\in\mathcal{A}\}} \alpha(X, Z) - g_X & \text{if } X \in \mathcal{S} \\ \sum_{\{Z:(X,Z)\in\mathcal{A}\}} \alpha(X, Z) & \text{if } X \in \mathcal{T} \\ \sum_{\{Z:(X,Z)\in\mathcal{A}\}} \alpha(X, Z) - g_X & \text{if } X \in \mathcal{R}. \end{cases} \tag{1.15}$$

[4] For example, there could be transshipment stages, i.e., $\mathcal{T} = \cup_{i=1}^{N} \mathcal{T}^i$ where points in $\mathcal{T}^i$ can ship and receive only to and from $\mathcal{T}^{i+1}$, $1 \le i \le N-1$, $\mathcal{T}^i$ can receive from $\mathcal{S}$, and $\mathcal{T}^i$ can ship to $\mathcal{R}$.

Notice that when $X \in \mathcal{R}$, there are no arcs $(X, Z) \in \mathcal{A}$ because all points in $\mathcal{R}$ are sinks, i.e., $v_X = -g_X$ for $X \in \mathcal{R}$. Notice also that since we choose all $\alpha(X, Y)$ sufficiently large for each arc $(X, Y) \in \mathcal{A}$, if follows that $v_X \geq 0$ for each $X \in \mathcal{N}$.

To have conservation, we need to verify that

$$\sum_{(X,Y)\in\mathcal{A}} \alpha(X, Y) = \sum_{X\in\mathcal{N}} v_X.$$

Indeed,

$$\begin{aligned}\sum_{X\in\mathcal{N}} v_X &= \sum_{X\in\mathcal{S}} \left(\sum_{\{Z:(X,Z)\in\mathcal{A}\}} \alpha(X, Z) - g_X \right) + \sum_{X\in\mathcal{T}} \left(\sum_{\{Z:(X,Z)\in\mathcal{A}\}} \alpha(X, Z) \right) \\ &\quad + \sum_{X\in\mathcal{R}} \left(\sum_{\{Z:(X,Z)\in\mathcal{A}\}} \alpha(X, Z) - g_X \right) \\ &= \sum_{(X,Z)\in\mathcal{A}} \alpha(X, Z) - \sum_{X\in\mathcal{S}} g_X - \sum_{X\in\mathcal{R}} g_X = \sum_{(X,Z)\in\mathcal{A}} \alpha(X, Z),\end{aligned}$$

since $\sum_{X\in\mathcal{S}} g_X + \sum_{X\in\mathcal{R}} g_X = 0$. We now define a cost function $c^* : \mathcal{A} \times \mathcal{N} \to [0, \infty)$ as follows. Let $c : \mathcal{N} \times \mathcal{N}$ be the original cost function of the network flow problem, denoted by N_0. For $(X, Y) \in \mathcal{A}$ and $Z \in \mathcal{N}$, we define

$$c^* ((X, Y); Z) = \begin{cases} 0 & \text{if } Z = X \\ c(X, Y) & \text{if } Z = Y \\ +\infty & \text{if } Z \neq X, Y \end{cases}$$

and then consider the optimal transport problem with the cost c^* and the capacities and demands defined before between the sets $\mathcal{A}$ and $\mathcal{N}$; let us denote this transport problem by T_0. Let $f ((X, Y); Z)$ be an admissible plan for T_0. This means

$$\sum_{Z\in\mathcal{N}} f ((X, Y); Z) = \alpha(X, Y) \qquad \forall (X, Y) \in \mathcal{A}$$

$$\sum_{(X,Y)\in\mathcal{A}} f ((X, Y); Z) = v_Z \qquad \forall Z \in \mathcal{N}$$

and since we seek to minimize

$$\sum_{(X,Y)\mathcal{A}, Z\in\mathcal{N}} f ((X, Y); Z)\, c^* ((X, Y); Z),$$

we must have $f((X,Y);Z) = 0$ for $Z \neq X, Y$ since $c^*((X,Y);Z) = \infty$ in that case. So the admissibility conditions read

$$\begin{cases} f((X,Y);X) + f((X,Y);Y) &= \alpha(X,Y) \quad \forall (X,Y) \in \mathcal{A} \\ \sum_{(X,Y)\in\mathcal{A}} f((X,Y);Z) &= v_Z \quad \forall Z \in \mathcal{N}. \end{cases} \tag{1.16}$$

Suppose $h(X,Y)$ is an admissible plan to the network flow problem N_0 with $0 \le h(X,Y) \le \alpha(X,Y)$ for each $(X,Y) \in \mathcal{A}$. If we set

$$f((X,Y);Z) = \begin{cases} h(X,Y) & \text{for } Z = Y \\ \alpha(X,Y) - h(X,Y) & \text{for } Z = X \\ 0 & \text{otherwise,} \end{cases} \tag{1.17}$$

then we shall prove that $f((X,Y);Z)$ is an admissible plan to the transportation problem T_0, i.e., $f((X,Y);Z)$ satisfies (1.16). In fact, since h satisfies

$$\sum_Y h(X,Y) - \sum_Y h(Y,X) = \begin{cases} g_X & \text{for } X \in \mathcal{S} \\ 0 & \text{for } X \in \mathcal{T} \\ g_X & \text{for } X \in \mathcal{R} \end{cases} \tag{1.18}$$

it follows that

$$\begin{aligned} \sum_{Z\in\mathcal{N}} f((X,Y);Z) &= \sum_{Z\in\mathcal{N}, Z=X,Y} f((X,Y);Z) = f((X,Y);X) + f((X,Y);Y) \\ &= \alpha(X,Y) - h(X,Y) + h(X,Y) = \alpha(X,Y) \quad \forall (X,Y) \in \mathcal{A}. \end{aligned}$$

Given $Z \in \mathcal{N}$, we write

$$\begin{aligned} \sum_{(X,Y)\in\mathcal{A}} f((X,Y);Z) &= \sum_{(X,Y)\in\mathcal{S}\times\mathcal{R}} f((X,Y);Z) + \sum_{(X,Y)\in\mathcal{S}\times\mathcal{T}} f((X,Y);Z) \\ &\quad + \sum_{(X,Y)\in\mathcal{T}\times\mathcal{T}\cap\mathcal{A}} f((X,Y);Z) + \sum_{(X,Y)\in\mathcal{T}\times\mathcal{R}} f((X,Y);Z) \\ &= A + B + C + D, \end{aligned} \tag{1.19}$$

and we shall prove $A + B + C + D = v_Z$.

Case $Z \in \mathcal{S}$ Then $C = D = 0$,

$$A = \sum_{(X,Y)\in\mathcal{S}\times\mathcal{R}} f((X,Y);Z) = \sum_{Y\in\mathcal{R}} f((Z,Y);Z) = \sum_{Y\in\mathcal{R}} (\alpha(Z,Y) - h(Z,Y)),$$

and

$$B = \sum_{(X,Y)\in\mathcal{S}\times\mathcal{T}} f\left((X,Y);Z\right) = \sum_{Y\in\mathcal{T}} f\left((Z,Y);Z\right) = \sum_{Y\in\mathcal{T}} \left(\alpha(Z,Y) - h(Z,Y)\right),$$

so

$$\begin{aligned}\sum_{(X,Y)\in\mathcal{A}} f\left((X,Y);Z\right) &= \sum_{Y\in\mathcal{R}\cup\mathcal{T}} \left(\alpha(Z,Y) - h(Z,Y)\right)\\ &= \sum_{Y\in\mathcal{R}\cup\mathcal{T}} \alpha(Z,Y) - \sum_{Y\in\mathcal{R}\cup\mathcal{T}} h(Z,Y)\\ &= v_Z + g_Z - \sum_{Y\in\mathcal{R}\cup\mathcal{T}} h(Z,Y) = v_Z \quad \text{since} \sum_{Y\in\mathcal{R}\cup\mathcal{T}} h(Y,Z)\\ &= 0 \text{ and } h \text{ is admissible.}\end{aligned}$$

Case $Z \in \mathcal{T}$ Then $A = 0$, and

$$B = \sum_{(X,Y)\in\mathcal{S}\times\mathcal{T}} f\left((X,Y);Z\right) = \sum_{X\in\mathcal{S}} f\left((X,Z);Z\right) = \sum_{X\in\mathcal{S}} h(X,Z).$$

Now

$$\begin{aligned}C &= \sum_{(X,Y)\in\mathcal{T}\times\mathcal{T}\cap\mathcal{A}} f\left((X,Y);Z\right)\\ &= \sum_{\{(X,Y)\in\mathcal{T}\times\mathcal{T}\cap\mathcal{A},X=Z \text{ or } Y=Z\}} f\left((X,Y);Z\right)\\ &\quad + \sum_{\{(X,Y)\in\mathcal{T}\times\mathcal{T}\cap\mathcal{A},X\neq Z \text{ and } Y\neq Z\}} f\left((X,Y);Z\right)\\ &= \sum_{\{(X,Y)\in\mathcal{T}\times\mathcal{T}\cap\mathcal{A},X=Z\}\cup\{(X,Y)\in\mathcal{T}\times\mathcal{T}\cap\mathcal{A},Y=Z\}} f\left((X,Y);Z\right)\\ &= \sum_{\{(X,Y)\in\mathcal{T}\times\mathcal{T}\cap\mathcal{A},X=Z\}} f\left((X,Y);Z\right) + \sum_{\{(X,Y)\in\mathcal{T}\times\mathcal{T}\cap\mathcal{A},Y=Z\}} f\left((X,Y);Z\right)\\ &= \sum_{\{Y\in\mathcal{T}:(Z,Y)\in\mathcal{A}\}} f\left((Z,Y);Z\right) + \sum_{\{X\in\mathcal{T}:(X,Z)\in\mathcal{A}\}} f\left((X,Z);Z\right)\\ &= \sum_{\{Y\in\mathcal{T}:(Z,Y)\in\mathcal{A}\}} \left(\alpha(Z,Y) - h(Z,Y)\right) + \sum_{\{X\in\mathcal{T}:(X,Z)\in\mathcal{A}\}} h(X,Z)\\ &= \sum_{\{Y\in\mathcal{T}:(Z,Y)\in\mathcal{A}\}} \alpha(Z,Y) + \sum_{\{X\in\mathcal{T}:(X,Z)\in\mathcal{A}\}} h(X,Z) - \sum_{\{X\in\mathcal{T}:(Z,X)\in\mathcal{A}\}} h(Z,X),\end{aligned}$$

and

$$D = \sum_{(X,Y)\in\mathcal{T}\times\mathcal{R}} f((X,Y);Z) = \sum_{(X,Y)\in\mathcal{T}\times\mathcal{R}, X=Z} f((X,Y);Z)$$
$$+ \sum_{(X,Y)\in\mathcal{T}\times\mathcal{R}, X\neq Z} f((X,Y);Z)$$
$$= \sum_{Y\in\mathcal{R}} f((Z,Y);Z) + 0 = \sum_{Y\in\mathcal{R}} (\alpha(Z,Y) - h(Z,Y)).$$

Hence

$$A + B + C + D = \sum_{X\in\mathcal{S}} h(X,Z) + \sum_{\{Y\in\mathcal{T}:(Z,Y)\in\mathcal{A}\}} \alpha(Z,Y) + \sum_{\{X\in\mathcal{T}:(X,Z)\in\mathcal{A}\}} h(X,Z)$$
$$- \sum_{\{X\in\mathcal{T}:(Z,X)\in\mathcal{A}\}} h(Z,X) + \sum_{Y\in\mathcal{R}} (\alpha(Z,Y) - h(Z,Y))$$
$$= \sum_{Y\in\{Y\in\mathcal{T}:(Z,Y)\in\mathcal{A}\}\cup\mathcal{R}} \alpha(Z,Y) + \sum_{X\in\{X\in\mathcal{T}:(X,Z)\in\mathcal{A}\}\cup\mathcal{S}} h(X,Z)$$
$$- \sum_{X\in\{X\in\mathcal{T}:(Z,X)\in\mathcal{A}\}\cup\mathcal{R}} h(Z,X)$$
$$= v_Z + 0 = v_Z, \quad \text{since } Z\in\mathcal{T} \text{ from (1.15) and (1.18).}$$

Case $Z \in \mathcal{R}$ We have

$$A = \sum_{X\in\mathcal{S}} f((X,Z);Z) = \sum_{X\in\mathcal{S}} h(X,Z);$$
$$B = C = 0;$$
$$D = \sum_{(X,Y)\in\mathcal{T}\times\mathcal{R}, Y=Z} f((X,Y);Z) + \sum_{(X,Y)\in\mathcal{T}\times\mathcal{R}, Y\neq Z} f((X,Y);Z)$$
$$= \sum_{X\in\mathcal{T}} f((X,Z);Z) + 0 = \sum_{X\in\mathcal{T}} h(X,Z).$$

Hence

$$A + B + C + D = \sum_{X\in\mathcal{S}\cup\mathcal{T}} h(X,Z) = \sum_{X\in\mathcal{N}} h(X,Z) \quad \text{since in } \mathcal{R} \text{ all are sinks}$$
$$= -g_Z + \sum_{X\in\mathcal{N}} h(Z,X) \quad \text{from (1.18)}$$
$$= v_Z \quad \text{from (1.15) since } h(Z,X) = 0 \text{ for each } Z \text{ sink.}$$

This concludes the verification that (1.17) is an admissible plan to the transportation problem T_0.

Reciprocally, if $f((X,Y);Z)$ is an admissible plan for T_0, and $h(X,Y) = f((X,Y);Y)$, then h is an admissible plan to the network problem N_0. In fact, we will verify that h satisfies (1.18) assuming $f((X,Y);Z)$ is an admissible plan, i.e., satisfies (1.16) and $f((X,Y);Z) = 0$ for $Z \neq X, Y$ since $c^*((X,Y);Z) = \infty$ in that case. We reverse the calculations done before.

Case When $Z \in \mathcal{S}$ We show that

$$\sum_Y h(Z,Y) - \sum_Y h(Y,Z) = g_Z. \tag{1.20}$$

From the definition of h

$$\begin{aligned}
&\sum_Y h(Z,Y) - \sum_Y h(Y,Z) \\
&= \sum_Y f((Z,Y);Y) - \sum_Y f((Y,Z);Z) \\
&= \sum_Y (\alpha(Z,Y) - f((Z,Y);Z)) - \sum_Y f((Y,Z);Z) \quad \text{from the 1st eq. in (1.16)} \\
&= \sum_Y \alpha(Z,Y) - \sum_Y (f((Z,Y);Z) + f((Y,Z);Z)) \\
&= v_Z + g_Z - \sum_Y (f((Z,Y);Z) + f((Y,Z);Z)) \quad \text{from (1.15).}
\end{aligned}$$

Now from (1.19) and since $Z \in \mathcal{S}$, we have

$$\begin{aligned}
v_Z &= \sum_{(X,Y)\in\mathcal{A}} f((X,Y);Z) = A + B = \sum_{Y\in\mathcal{R}} f((Z,Y);Z) + \sum_{Y\in\mathcal{T}} f((Z,Y);Z) \\
&= \sum_Y f((Z,Y);Z).
\end{aligned}$$

Since $Z \in \mathcal{S}$, Z is a source, and so there are no arcs (Y,Z) with $Y \in \mathcal{N}$, so $\sum_Y f((Y,Z);Z) = 0$. Therefore (1.20) follows.

Case When $Z \in \mathcal{T}$ We show that

$$\sum_Y h(Z,Y) - \sum_Y h(Y,Z) = 0. \tag{1.21}$$

From the definition of h

$$\begin{aligned}
&\sum_Y h(Z,Y) - \sum_Y h(Y,Z)\\
&= \sum_Y f((Z,Y);Y) - \sum_Y f((Y,Z);Z)\\
&= \sum_Y (\alpha(Z,Y) - f((Z,Y);Z)) - \sum_Y f((Y,Z);Z) \quad \text{from the 1st eq. in (1.16)}\\
&= \sum_Y \alpha(Z,Y) - \sum_Y (f((Z,Y);Z) + f((Y,Z);Z))\\
&= v_Z - \sum_Y (f((Z,Y);Z) + f((Y,Z);Z)) \quad \text{from (1.15).}
\end{aligned}$$

Now from (1.19) and since $Z \in \mathcal{T}$, we have

$$\begin{aligned}
v_Z &= \sum_{(X,Y)\in\mathcal{A}} f((X,Y);Z) = B + C + D\\
&= \sum_{X\in\mathcal{S}} f((X,Z);Z) + \sum_{Y\in\mathcal{T}:(Z,Y)\in\mathcal{A}} f((Z,Y);Z) + \sum_{X\in\mathcal{T}:(X,Z)\in\mathcal{A}} f((X,Z);Z)\\
&\quad + \sum_{Y\in\mathcal{R}} f((Z,Y);Z)\\
&= \sum_{X\in\mathcal{S}\cup\mathcal{T}:(X,Z)\in\mathcal{A}} f((X,Z);Z) + \sum_{Y\in\mathcal{R}\cup\mathcal{T}:(Z,Y)\in\mathcal{A}} f((Z,Y);Z)\\
&= \sum_{X\in\mathcal{N}} f((X,Z);Z) + \sum_{Y\in\mathcal{N}} f((Z,Y);Z) \quad \text{since } \mathcal{R} \text{ are only sinks and } \mathcal{S}\\
&\quad \text{are only sources}
\end{aligned}$$

and (1.21) follows.

Case When $Z \in \mathcal{R}$ We show that

$$\sum_Y h(Z,Y) - \sum_Y h(Y,Z) = g_Z. \tag{1.22}$$

From the definition of h

$$\begin{aligned}
&\sum_Y h(Z,Y) - \sum_Y h(Y,Z)\\
&= \sum_Y f((Z,Y);Y) - \sum_Y f((Y,Z);Z)
\end{aligned}$$

$$= \sum_{Y} (\alpha(Z, Y) - f((Z, Y); Z)) - \sum_{Y} f((Y, Z); Z) \quad \text{from the 1st eq. in (1.16)}$$

$$= \sum_{Y} \alpha(Z, Y) - \sum_{Y} (f((Z, Y); Z) + f((Y, Z); Z))$$

$$= v_Z + g_Z - \sum_{Y} (f((Z, Y); Z) + f((Y, Z); Z)) \quad \text{from (1.15).}$$

Now from (1.19) and since $Z \in \mathcal{R}$, we have

$$v_Z = \sum_{(X,Y) \in \mathcal{A}} f((X, Y); Z) = A + D = \sum_{Y \in \mathcal{S}} f((Y, Z); Z) + \sum_{Y \in \mathcal{T}} f((Y, Z); Z)$$

$$= \sum_{Y} f((Y, Z); Z),$$

since $\mathcal{R}$ are sinks. Finally, $\sum_Y f((Z, Y); Z) = 0$ since $\mathcal{R}$ are sinks and so there are no arcs (Z, Y) for all $Y \in \mathcal{N}$. Therefore (1.22) follows.

So there is a one to one correspondence between admissible plans of T_0 and N_0, and in addition, from the definition of c^*, corresponding admissible plans have the same cost. Therefore the minimum cost of N_0 equals the minimum cost of T_0.

1.8 More on the Transshipment Problem

Let (X_i, d_i) be metric spaces with finite diameter, $\text{diam}(X_i) = \Delta_i < \infty$, $i = 1, 2$. Suppose X_1, X_2 are disjoint and let $\mathcal{N} = X_1 \cup X_2$. Define

$$d(x, y) = \begin{cases} d_1(x, y) & \text{if } x, y \in X_1 \\ d_2(x, y) & \text{if } x, y \in X_2 \\ \max\{\Delta_1, \Delta_2\} & \text{if } x \in X_1, y \in X_2 \text{ or } x \in X_2, y \in X_1. \end{cases}$$

It is easy to see that d is a distance in $\mathcal{N}$ and $\text{diam}(\mathcal{N}) < \infty$.

Let Σ_i be the Borel σ-algebra of subsets of X_i, $i = 1, 2$. The class

$$\Sigma = \{A \cup B : A \in \Sigma_1, B \in \Sigma_2\},$$

is the Borel σ-algebra of subsets of $\mathcal{N}$. Let μ_i be a finite Borel signed measure on X_i, $i = 1, 2$. If we define

$$\mu(E) = \mu_1(A) + \mu_2(B) \qquad \text{for } E \in \Sigma, E = A \cup B,$$

then it is easy to see that μ is a finite Borel signed measure in $\mathcal{N}$.

Let $c : \mathcal{N} \times \mathcal{N} \to [0, \infty)$ be a Borel measurable function and μ_i be finite Borel measures on X_i, $i = 1, 2$, with $\mu_1(X_1) = \mu_2(X_2)$. Then the transshipment problem is to minimize

$$\int_{\mathcal{N}\times\mathcal{N}} c(x, y)\, d\gamma(x, y)$$

over all Borel measures γ in $\mathcal{N} \times \mathcal{N}$ satisfying

$$\gamma(E \times \mathcal{N}) - \gamma(\mathcal{N} \times E) = \mu_1(A_i) - \mu_2(A_2), \quad \forall E = A_1 \cup A_2; \quad A_i \in \Sigma_i.$$

1.9 Linear Programming

Let $c \in \mathbb{R}^N$, $b \in \mathbb{R}^M$, $A \in \mathbb{R}^{M\times N}$. A linear programming problem has the form

$$\text{minimize} \quad x \cdot c \qquad \text{over } x \in \mathbb{R}^N \text{ in the set} Ax = b, \quad x \geq 0. \tag{1.23}$$

This is called *the primal problem.*[5] It has associated the so called *dual problem* defined as follows:

$$\text{maximize} \quad y \cdot b \qquad \text{over } y \in \mathbb{R}^M \text{ in the set } A^t y \leq c. \tag{1.24}$$

see [12, pp. 61–62 and pp.124–144].

We now find the dual problem for the transportation problem described in Sect. 1.1. We first need to write the transportation problem in the form (1.23), i.e., we need to rewrite the constraints

$$\sum_{j=1}^{n} x_{ij} = u_i, 1 \leq i \leq m; \qquad \sum_{i=1}^{m} x_{ij} = v_j, 1 \leq j \leq n.$$

Let us set

$$x = (x_{11}, x_{12}, \cdots, x_{1n}, x_{21}, \cdots, x_{2n}, x_{31}, \cdots, x_{3n}, \cdots, x_{m1}, \cdots, x_{mn}) \tag{1.25}$$

[5] If the equality $Ax = b$ is replaced by the inequality $Ax \geq b$, then in the dual problem one needs to assume $y \geq 0$; see [12, p. 126] for various options to define the primal and dual problems.

so $x \in \mathbb{R}^{mn}$. Now let I_n be the identity matrix in $\mathbb{R}^{n\times n}$, $1_n = (1, 1, \cdots, 1) \in \mathbb{R}^n$, and define

$$A = \begin{pmatrix} I_n & I_n & I_n & \cdots & I_n \\ 1_n & 0 & 0 & \cdots & 0 \\ 0 & 1_n & 0 & \cdots & 0 \\ \vdots & \vdots & \vdots & \cdots & \vdots \\ 0 & 0 & 0 & \cdots & 1_n \end{pmatrix}$$

where the matrix I_n appear m times and also 1_n appears m times. So $A \in \mathbb{R}^{(n+m)\times(mn)}$, and from (1.25) the constraints in (1.23) can be written as

$$Ax = b$$

where $b = (v, u) \in \mathbb{R}^{n+m}$ with $v = (v_1, \cdots, v_n)$ and $u = (u_1, \cdots, u_m)$. Now

$$A^t = \begin{pmatrix} I_n & 1_n^t & 0 & \cdots & 0 \\ I_n & 0 & 1_n^t & \cdots & 0 \\ I_n & 0 & 0 & \cdots & 0 \\ \vdots & \vdots & \vdots & \cdots & \vdots \\ I_n & 0 & 0 & \cdots & 1_n^t \end{pmatrix},$$

$A^t \in \mathbb{R}^{(mn)\times(n+m)}$ where 1_n^t is the column vector with its n components equal one. If we set $y = (y^1, y^2)$ with $y^1 = \left(y_1^1, \cdots, y_n^1\right) \in \mathbb{R}^n$ and $y^2 = \left(y_1^2, \cdots, y_m^2\right) \in \mathbb{R}^m$, then $A^t y \le c$ reads

$$y_i^1 + y_j^2 \le c_{ij} \tag{1.26}$$

for $1 \le i \le n$ and $1 \le j \le m$, and so the dual of the transportation problem is to maximize

$$y^1 \cdot v + y^2 \cdot u$$

over all $y^1 \in \mathbb{R}^n$ and $y^2 \in \mathbb{R}^m$ satisfying (1.26).

The dual problem considered here is the discrete version of the general dual problem analyzed Chap. 6.

Remark 1.7 We shall prove what is called the weak duality theorem. Let us define the feasible sets for each the primal and dual problems:

$$F_p = \{x \in \mathbb{R}^N : x \ge 0, Ax = b\}$$

$$F_d = \{y \in \mathbb{R}^M : A^t y \le c\}.$$

If $x \in F_p$ and $y \in F_d$, then

$$b \cdot y = (Ax) \cdot y = x \cdot A^t y \leq x \cdot c,$$

so

$$\sup_{y \in F_d} y \cdot b \leq \inf_{x \in F_p} x \cdot c.$$

The reverse inequality holds true under certain conditions and is called the strong duality theorem, see [12, p. 129]. This will be proved later in a more general case, see Remark 6.8.

Chapter 2
The Normal Mapping or Subdifferential

Abstract This chapter describes basic properties of the normal mapping or subdifferential and introduces the notion of Monge-Ampère measure and its properties. The chapter concludes with a list of exercises helping to understand these concepts.

In this chapter, we describe basic properties of the subdifferential and the Monge-Ampère measure. For further results and applications we refer to [32].

Let Ω be an open subset of $\mathbb{R}^n$ and $u : \Omega \to \mathbb{R}$. Given $x_0 \in \Omega$, *a supporting hyperplane to the function u at the point* $(x_0, u(x_0))$ is an affine function $\ell(x) = u(x_0) + p \cdot (x - x_0)$ such that $u(x) \geq \ell(x)$ for all $x \in \Omega$.

Definition 2.1 The normal mapping of u, or subdifferential of u, is the set-valued function $\partial u : \Omega \to \mathcal{P}(\mathbb{R}^n)$ defined by

$$\partial u(x_0) = \{p : u(x) \geq u(x_0) + p \cdot (x - x_0), \quad \text{for all } x \in \Omega\}.$$

Given $E \subset \Omega$, we define $\partial u(E) = \bigcup_{x \in E} \partial u(x)$.

The set $\partial u(x_0)$ may be empty. Let $S = \{x \in \Omega : \partial u(x) \neq \emptyset\}$. If $u \in C^1(\Omega)$ and $x \in S$, then $\partial u(x) = Du(x)$, the gradient of u at x, which means that when u is differentiable the normal mapping is basically the gradient; see Exercise 11. If $u \in C^2(\Omega)$ and $x \in S$, then the Hessian of u is nonnegative definite, that is $D^2u(x) \geq 0$. This means that if u is C^2, then S is the set where the graph of u is concave up. Indeed, by Taylor's Theorem $u(x+h) = u(x) + Du(x) \cdot h + \frac{1}{2}\langle D^2u(\xi)h, h\rangle$, where ξ lies on the segment between x and $x + h$. Since $u(x + h) \geq u(x) + Du(x) \cdot h$ for all h sufficiently small, the claim follows.

Example 2.2 It is useful to calculate the normal mapping of the function u whose graph is a cone in $\mathbb{R}^{n+1}$. Let $\Omega = B_R(x_0)$ in $\mathbb{R}^n$, $h > 0$ and $u(x) = h\,\frac{|x - x_0|}{R}$. The graph of u, for $x \in \Omega$, is an upside-down right-cone in $\mathbb{R}^{n+1}$ with vertex at the

C. E. Gutiérrez, *Optimal Transport and Applications to Geometric Optics*,
SpringerBriefs on PDEs and Data Science,
https://doi.org/10.1007/978-981-99-4867-3_2

point $(x_0, 0)$ and base on the hyperplane $x_{n+1} = h$. We shall show that

$$\partial u(x) = \begin{cases} \dfrac{h}{R}\dfrac{x - x_0}{|x - x_0|}, & \text{for } 0 < |x - x_0| < R, \\ \\ \overline{B_{h/R}(0)}, & \text{for } x = x_0. \end{cases}$$

If $0 < |x - x_0| < R$, then the value of ∂u follows by calculating the gradient. By the definition of normal mapping, $p \in \partial u(x_0)$ if and only if $\frac{h}{R}|x - x_0| \geq p \cdot (x - x_0)$ for all $x \in B_R(x_0)$. If $p \neq 0$ and we pick $x = x_0 + R\frac{p}{|p|}$, then $|p| \leq \frac{h}{R}$. It is clear that $|p| \leq \frac{h}{R}$ implies $p \in \partial u(x_0)$.

2.1 Properties of the Normal Mapping

Lemma 2.2 *If $\Omega \subset \mathbb{R}^n$ is open, $u \in C(\Omega)$ and $K \subset \Omega$ is compact, then $\partial u(K)$ is compact.*

Proof Let $\{p_k\} \subset \partial u(K)$ be a sequence. We claim that p_k is bounded. For each k there exists $x_k \in K$ such that $p_k \in \partial u(x_k)$, that is $u(x) \geq u(x_k) + p_k \cdot (x - x_k)$ for all $x \in \Omega$. Since K is compact, $K_\delta = \{x : \text{dist}(x, K) \leq \delta\}$ is compact and contained in Ω for all δ sufficiently small, and we may assume by passing if necessary through a subsequence that $x_k \to x_0$. Then $x_k + \delta w \in K_\delta$, and $u(x_k + \delta w) \geq u(x_k) + \delta p_k \cdot w$ for all $|w| = 1$ and for all k. If $p_k \neq 0$ and $w = \frac{p_k}{|p_k|}$ then we get $\max_{K_\delta} u(x) \geq \min_K u(x) + \delta |p_k|$, for all k. Since u is locally bounded, the claim is proved. Hence there exists a convergent subsequence $p_{k_m} \to p_0$. We claim that $p_0 \in \partial u(K)$. We shall prove that $p_0 \in \partial u(x_0)$. We have $u(x) \geq u(x_{k_m}) + p_{k_m} \cdot (x - x_{k_m})$ for all $x \in \Omega$ and, since u is continuous, by letting $m \to \infty$ we obtain $u(x) \geq u(x_0) + p_0 \cdot (x - x_0)$ for all $x \in \Omega$. This completes the proof of the lemma. □

Remark 2.4 We note that the proof above shows that if u is only locally bounded in Ω, then $\partial u(E)$ is bounded whenever E is bounded with $\overline{E} \subset \Omega$.

Remark 2.5 We note that given $x_0 \in \Omega$, the set $\partial u(x_0)$ is convex. However, if K is convex and $K \subset \Omega$, then the set $\partial u(K)$ is not necessarily convex. An example is given by $u(x) = e^{|x|^2}$ and $K = \{x \in \mathbb{R}^n : |x_i| \leq 1, \quad i = 1, \ldots, n\}$. The set $\partial u(K)$ is a star-shaped symmetric set around the origin that is not convex, see Fig. 2.1.

Lemma 2.3 *If u is a convex function in Ω and $K \subset \Omega$ is compact, then u is uniformly Lipschitz in K, that is, there exists a constant $C = C(u, K)$ such that $|u(x) - u(y)| \leq C|x - y|$ for all $x, y \in K$.*

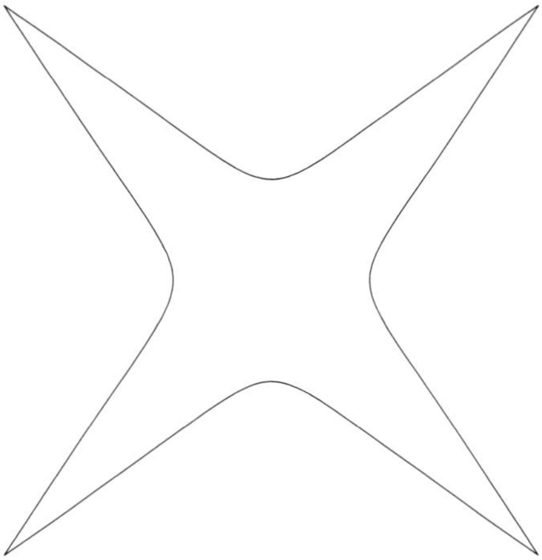

Fig. 2.1 $\partial u(K)$

Proof Since u is convex, u has a supporting hyperplane at any $x \in \Omega$. Let $C = \sup\{|p| : p \in \partial u(K)\}$. By Lemma 2.2, $C < \infty$. If $x \in K$, then $u(y) \geq u(x) + p \cdot (y-x)$ for $p \in \partial u(x)$ and for all $y \in \Omega$. In particular, if $y \in K$, then $u(y) - u(x) \geq -|p||y-x|$. By reversing the roles of x and y we get the lemma.

□

Lemma 2.4 *If Ω is open and u is Lipschitz continuous in Ω, then u is differentiable a.e. in Ω.*

Proof See [15, p. 81]. □

Lemma 2.5 *If u is convex or concave in Ω, then u is differentiable a.e. in Ω.*

Proof Follows immediately from Lemmas 2.3 and 2.4. □

Remark 2.9 A deep result of Busemann–Feller–Aleksandrov establishes that any convex function in Ω has second order derivatives a.e., see [15, p. 242] and [52, pp. 31–32].

Definition 2.6 The Legendre transform of the function $u : \Omega \to \mathbb{R}$ is the function $u^* : \mathbb{R}^n \to \mathbb{R}$ defined by

$$u^*(p) = \sup_{x \in \Omega} (x \cdot p - u(x)) .$$

Remark 2.11 If Ω is bounded and u is bounded in Ω, then u^* is finite. Also, u^* is convex in $\mathbb{R}^n$.

Lemma 2.7 *If Ω is open and u is a continuous function in Ω, then the set of points in $\mathbb{R}^n$ that belong to the image by the normal mapping of more than one point of Ω has Lebesgue measure zero. That is, the set*

$$S = \{p \in \mathbb{R}^n : \textit{there exist } x, y \in \Omega, x \neq y \textit{ and } p \in \partial u(x) \cap \partial u(y)\}$$

has measure zero. This also means that the set of supporting hyperplanes that touch the graph of u at more than one point has measure zero.

Proof We may assume that Ω is bounded because otherwise we write $\Omega = \cup_k \Omega_k$, where $\Omega_k \subset \Omega_{k+1}$ are open and $\overline{\Omega_k}$ are compact. If $p \in S$, then there exist $x, y \in \Omega$, $x \neq y$ and $u(z) \geq u(x) + p \cdot (z - x)$, $u(z) \geq u(y) + p \cdot (z - y)$ for all $z \in \Omega$. Since Ω_k increases, $x, y \in \Omega_m$ for some m and obviously the previous inequalities hold true for $z \in \Omega_m$. That is, if

$$S_m = \{p \in \mathbb{R}^n : \text{there exist } x, y \in \Omega, x \neq y \text{ and } p \in \partial(u|\Omega_m)(x) \cap \partial(u|\Omega_m)(y)\}$$

we have $p \in S_m$, i.e., $S \subset \cup_m S_m$ and we then show that each S_m has measure zero.

Let u^* be the Legendre transform of u. By Remark 2.11 and Lemma 2.5, u^* is differentiable a.e. Let $E = \{p : u^* \text{ is not differentiable at } p\}$. We shall show that

$$\{p \in \mathbb{R}^n : \text{there exist } x, y \in \Omega, x \neq y \text{ and } p \in \partial u(x) \cap \partial u(y)\} \subset E.$$

In fact, if $p \in \partial u(x_1) \cap \partial u(x_2)$ and $x_1 \neq x_2$, then $u^*(p) = x_i \cdot p - u(x_i)$, $i = 1, 2$. Also $u^*(z) \geq x_i \cdot z - u(x_i)$ and so $u^*(z) \geq u^*(p) + x_i \cdot (z - p)$ for all z, $i = 1, 2$. Hence if u^* were differentiable at p we would have $Du^*(p) = x_i$, $i = 1, 2$. This completes the proof of the lemma. □

Theorem 2.13 *If Ω is open and $u \in C(\Omega)$, then the class*

$$\mathcal{S} = \{E \subset \Omega : \partial u(E) \text{ is Lebesgue measurable}\}$$

is a Borel σ-algebra. The set function $Mu : \mathcal{S} \to \overline{\mathbb{R}}$ defined by

$$Mu(E) = |\partial u(E)| \tag{2.1}$$

is a measure, finite on compacts, that is called the Monge–Ampère measure associated with the function u.

Proof By Lemma 2.2, the class $\mathcal{S}$ contains all compact subsets of Ω. Also, if E_m is any sequence of subsets of Ω, then $\partial u\left(\cup_m E_m\right) = \cup_m \partial u(E_m)$. Hence, if $E_m \in \mathcal{S}$, $m = 1, 2, \dots$, then $\cup_m E_m \in \mathcal{S}$. In particular, we may write $\Omega = \cup_m K_m$ with K_m compacts and we obtain $\Omega \in \mathcal{S}$. To show that $\mathcal{S}$ is a σ-algebra it remains to show that if $E \in \mathcal{S}$, then $\Omega \setminus E \in \mathcal{S}$. We use the following formula, which is valid for any set $E \subset \Omega$:

$$\partial u(\Omega \setminus E) = (\partial u(\Omega) \setminus \partial u(E)) \cup (\partial u(\Omega \setminus E) \cap \partial u(E)). \tag{2.2}$$

By Lemma 2.7, $|\partial u(\Omega \setminus E) \cap \partial u(E)| = 0$ for any set E. Then from (2.2) we get $\Omega \setminus E \in \mathcal{S}$ when $E \in \mathcal{S}$.

We now show that Mu is σ-additive. Let $\{E_i\}_{i=1}^{\infty}$ be a sequence of disjoint sets in $\mathcal{S}$ and set $\partial u(E_i) = H_i$. We must show that

$$\left|\partial u\left(\cup_{i=1}^{\infty} E_i\right)\right| = \sum_{i=1}^{\infty} |H_i|.$$

Since $\partial u\left(\cup_{i=1}^{\infty} E_i\right) = \cup_{i=1}^{\infty} H_i$, we shall show that

$$\left|\cup_{i=1}^{\infty} H_i\right| = \sum_{i=1}^{\infty} |H_i|. \tag{2.3}$$

We have $E_i \cap E_j = \emptyset$ for $i \neq j$. Then by Lemma 2.7 $|H_i \cap H_j| = 0$ for $i \neq j$. Let us write

$$\cup_{i=1}^{\infty} H_i = H_1 \cup (H_2 \setminus H_1) \cup (H_3 \setminus (H_2 \cup H_1)) \cup (H_4 \setminus (H_3 \cup H_2 \cup H_1)) \cup \cdots,$$

where the sets on the right hand side are disjoint. Now

$$H_n = [H_n \cap (H_{n-1} \cup H_{n-2} \cup \cdots \cup H_1)] \cup [H_n \setminus (H_{n-1} \cup H_{n-2} \cup \cdots \cup H_1)].$$

Then by Lemma 2.7, $|H_n \cap (H_{n-1} \cup H_{n-2} \cup \cdots \cup H_1)| = 0$ and we obtain

$$|H_n| = |H_n \setminus (H_{n-1} \cup H_{n-2} \cup \cdots \cup H_1)|.$$

Consequently (2.3) follows, and the proof of the theorem is complete. □

Example 2.3 If $u \in C^2(\Omega)$ is a convex function, then the Monge–Ampère measure Mu associated with u satisfies

$$Mu(E) = \int_E \det D^2 u(x)\, dx, \tag{2.4}$$

for all Borel sets $E \subset \Omega$. To prove (2.4), we use the following result:

Theorem 2.15 (Sard's Theorem, see [42]) *Let $\Omega \subset \mathbb{R}^n$ be an open set and $g : \Omega \to \mathbb{R}^n$ a C^1 function in Ω. If $S_0 = \{x \in \Omega : \det g'(x) = 0\}$, then $|g(S_0)| = 0$.*

We first notice that since u is convex and $C^2(\Omega)$, then Du is one-to-one on the set $A = \{x \in \Omega : D^2u(x) > 0\}$. Indeed, let $x_1, x_2 \in A$ with $Du(x_1) = Du(x_2)$. By convexity $u(z) \geq u(x_i) + Du(x_i) \cdot (z - x_i)$ for all $z \in \Omega$, $i = 1, 2$. Hence $u(x_1) - u(x_2) = Du(x_1) \cdot (x_1 - x_2) = Du(x_2) \cdot (x_1 - x_2)$. By the Taylor formula we can write

$$u(x_1) = u(x_2) + Du(x_2) \cdot (x_1 - x_2) \\ + \int_0^1 t\langle D^2u\,(x_2 + t(x_1 - x_2))\,(x_1 - x_2), x_1 - x_2\rangle\,dt.$$

Therefore the integral is zero and the integrand must vanish for $0 \le t \le 1$. Since $x_2 \in A$, it follows that $x_2 + t(x_1 - x_2) \in A$ for t small. Therefore $x_1 = x_2$.

If $u \in C^2(\Omega)$, then $g = Du \in C^1(\Omega)$. We have $Mu(E) = |Du(E)|$ and

$$Du(E) = Du(E \cap S_0) \cup Du(E \setminus S_0).$$

Since $E \subset \mathbb{R}^n$ is a Borel set, $E \cap S_0$ and $E \setminus S_0$ are also Borel sets. Hence, by the formula of change of variables and Sard's Theorem,

$$Mu(E) = Mu(E \cap S_0) + Mu(E \setminus S_0) = \int_{E\setminus S_0} \det D^2u(x)\,dx = \int_E \det D^2u(x)\,dx,$$

which shows (2.4).

Example 2.16 If $u(x)$ is the cone of Example 2.2, then the Monge–Ampère measure associated with u is $Mu = |B_{h/R}|\,\delta_{x_0}$, where δ_{x_0} denotes the Dirac delta at x_0.

2.2 Weak Convergence of Monge-Ampère Measures

Here we show that the notion of Monge-Ampère measure is stable by uniform limits. That is, if $u_k \to u$ uniformly on compact subsets of Ω, then $Mu_k \to Mu$ weakly. First we prove the following lemma.

Lemma 2.16 *Let $u_k \in C(\Omega)$ be convex functions such that $u_k \to u$ uniformly on compact subsets of Ω.*

We have

(i) If $K \subset \Omega$ is compact, then

$$\limsup_{k\to\infty} \partial u_k(K) \subseteq \partial u(K),$$

and by Fatou's lemma

$$\limsup_{k\to\infty} |\partial u_k(K)| \le |\partial u(K)|.$$

(ii) If U is open such that $U \subset \Omega$, then

$$\partial u(U) \subseteq \liminf_{k\to\infty} \partial u_k(U),$$

where the inequality holds for almost every point of the set on the left-hand side,[1] *and by Fatou's lemma*

$$|\partial u(U)| \leq \liminf_{k\to\infty} |\partial u_k(U)|.$$

Proof

(i) If $p \in \limsup_{k\to\infty} \partial u_k(K)$, then for each n there exist k_n and $x_{k_n} \in K$ such that $p \in \partial u_{k_n}(x_{k_n})$. By selecting a subsequence x_j of x_{k_n} we may assume that $x_j \to x_0 \in K$. On the other hand,

$$u_j(x) \geq u_j(x_j) + p \cdot (x - x_j), \qquad \forall x \in \Omega,$$

and by letting $j \to \infty$, by the uniform convergence of u_j on compacts we get

$$u(x) \geq u(x_0) + p \cdot (x - x_0), \qquad \forall x \in \Omega,$$

that is $p \in \partial u(x_0)$.

(ii) Let $S = \{p : p \in \partial u(x_1) \cap \partial u(x_2) \quad \text{for some } x_1, x_2 \in \Omega, x_1 \neq x_2\}$. By Lemma 2.7, $|S| = 0$. Let $U \subset \Omega$ be open and consider $\partial u(U) \setminus S$. If $p \in \partial u(U) \setminus S$, then there exists a unique $x_0 \in U$ such that $p \in \partial u(x_0)$ and $p \notin \partial u(x_1)$ for all $x_1 \in \Omega$, $x_1 \neq x_0$. If $x_1 \in \Omega$ and $x_1 \neq x_0$, then $u(x_1) > u(x_0) + p \cdot (x_1 - x_0)$. Otherwise, $u(x_1) = u(x_0) + p \cdot (x_1 - x_0)$ and since $p \in \partial u(x_0)$ we have

$$\begin{aligned} u(x) &\geq u(x_0) + p \cdot (x - x_0) \qquad \forall x \in \Omega \\ &= u(x_1) - p \cdot (x_1 - x_0) + p \cdot (x - x_0) \\ &= u(x_1) + p \cdot (x - x_1) \qquad \forall x \in \Omega, \end{aligned}$$

that is, $p \in \partial u(x_1)$ which is impossible because we removed S from $\partial u(U)$. Let us first assume that $\overline{U}$ is compact, let $\ell(x) = u(x_0) + p \cdot (x - x_0)$, and set $\delta = \min\{u(x) - \ell(x) : x \in \partial U\} > 0$. From the uniform convergence we have that $|u(x) - u_k(x)| < \delta/2$ for all $x \in \overline{U}$ and for all $k \geq k_0$. Let

$$\delta_k = \max_{x\in\overline{U}}\{\ell(x) - u_k(x) + \delta/2\}.$$

[1] The inclusion holds for $\partial u(U) \setminus S$, where $S = \{p : p \in \partial u(x_1) \cap \partial u(x_2) \quad \text{for some } x_1, x_2 \in \Omega, x_1 \neq x_2\}$.

Since $x_0 \in U$, we have $\delta_k \geq \ell(x_0) - u_k(x_0) + \delta/2 = u(x_0) - u_k(x_0) + \delta/2 > -\delta/2 + \delta/2 = 0$. We have $\delta_k = \ell(x_k) - u_k(x_k) + \delta/2$ for some $x_k \in \overline{U}$. We claim that $x_k \notin \partial U$. Otherwise, by definition of δ, $\ell(x_k) - u(x_k) \leq -\delta$ and so $\delta_k \leq -\delta/2$, a contradiction. We claim that p is the slope of a supporting hyperplane to u_k at the point $(x_k, u(x_k))$. Indeed,

$$\delta_k = u(x_0) + p \cdot (x_k - x_0) - u_k(x_k) + \delta/2$$

and so

$$u_k(x) \geq u_k(x_k) + p \cdot (x - x_k) \qquad \forall x \in \overline{U}. \tag{2.5}$$

Since u_k is convex in Ω, U is open, and $x_k \in U$, it follows that (2.5) holds for all $x \in \Omega$, that is $p \in \partial u_k(x_k)$ for all $k \geq k_0$. This implies that $p \in \liminf_{k\to\infty} \partial u_k(U)$.

Finally, we remove the assumption that $\overline{U}$ is compact. If U is open, and $U \subset \Omega$, then we can write $U = \bigcup_{j=1}^{\infty} U_j$ with U_j open and $\overline{U_j}$ compact. Then

$$\begin{aligned}\partial u(U) &= \bigcup_{j=1}^{\infty} \partial u(U_j) \subseteq \bigcup_{j=1}^{\infty} \liminf_{k\to\infty} \partial u_k(U_j) \subseteq \bigcup_{j=1}^{\infty} \liminf_{k\to\infty} \partial u_k(U)\\ &= \liminf_{k\to\infty} \partial u_k(U),\end{aligned}$$

which completes the proof.

□

Lemma 2.17 *If u_k are convex functions in Ω such that $u_k \to u$ uniformly on compact subsets of Ω, then u is convex and the associated Monge–Ampère measures Mu_k tend to Mu weakly, that is*

$$\int_{\Omega} f(x)\, dMu_k(x) \to \int_{\Omega} f(x)\, dMu(x),$$

for every f continuous with compact support in Ω.

See Exercise 20 for a proof.

2.3 Exercises on the Subdifferential and Monge-Ampère Measures

(1) Given $u : \Omega \to \mathbb{R}$, the local subdifferential of u is given by

$$\partial_\ell u(y) = \{p \in \mathbb{R}^n : \text{there exists a neighborhood } U_y \text{ of } y \text{ such that}$$
$$u(x) \geq u(y) + p \cdot (x - y)\, \forall x \in U_y\}.$$

Prove that:

(a) $\partial u(y) \subset \partial_\ell u(y)$ for each $y \in \Omega$, where the inequality may be strict.
(b) if Ω is convex and u is convex in Ω, then $\partial_\ell(y) = \partial u(y)$ for each $y \in \Omega$.
(c) if $u : \Omega \to \mathbb{R}$ is locally convex in Ω convex, i.e., for each $x \in \Omega$ there exists $\delta > 0$ and an affine function ℓ_x such that $\ell_x(z) \leq u(z)$ for all $|z - x| < \delta$, then u is convex in Ω.
HINT: Assume first the strict convexity assumption: for each $x \in \Omega$ there exists $\delta > 0$ and an affine function ℓ_x such that $\ell_x(z) < u(z)$ for all $|z - x| < \delta$ with $z \neq x$ and with equality at $z = x$. Proceed by contradiction: there exist $a, b \in \Omega$ such that $u(ta + (1-t)b) > tu(a) + (1-t)u(b)$ for some $0 < t < 1$. Let $M = \max_{0 \leq t \leq 1} u(ta+(1-t)b) - tu(a) - (1-t)u(b)$. Then $M > 0$ and the maximum is attained at some $0 < s < 1$. Let $x_s = sa+(1-s)b$. Consider the affine function ℓ_{x_s} such that $\ell_{x_s}(z) < u(z)$ for all $|z - x| < \delta$, $z \neq x$ with equality at $z = x$, and show that M is not maximum.
Second: Take $v(x) = u(x) + \epsilon|x|^2$ with $\epsilon > 0$. Since $\epsilon|x|^2$ is strictly convex, then v verifies the strict convexity assumption and so v is convex for all $\epsilon > 0$. Letting $\epsilon \to 0$ we obtain the result.

(2) Let $F(A) = \det A$ for each $n \times n$ symmetric matrix A. Prove that $\dfrac{\partial F}{\partial a_{ij}}(A) = A^{ij}$ where $\left(A^{ij}\right)_{ij}$ is the cofactor matrix of A.

(3) The Monge-Ampère operator is of divergence form. If u is C^3 and U^{ij} is the cofactor matrix of D^2u, then prove that $\det D^2 u = \dfrac{1}{n} U^{ij} D_{ij} u = \dfrac{1}{n} D_i \left(U^{ij} D_j u\right)$, where U^{ij} is divergence free, i.e., $\sum_{i=1}^n D_i U^{ij} = 0$ for $1 \leq j \leq n$.

(4) Let A be symmetric and non negative definite. Prove that

$$(\det A)^{1/n} = \inf \left\{ \frac{1}{n} \operatorname{trace}(AB) : B \text{ is symmetric and } \det B = 1 \right\}.$$

HINT: since B is symmetric and positive definite, then $B = O^t D O$ with O orthogonal and D diagonal with diagonal λ_i. Since $\operatorname{trace}(AB) = \operatorname{trace}(BA)$ for all matrices A, B, we have $\operatorname{trace}(AB) = \operatorname{trace}(O^t A O D) = \sum_{i=1}^n a'_{ii} \lambda_i$

where a'_{ii} is the diagonal of $O^t AO$. Next use the arithmetic-geometric inequality. To estimate $a'_{11} \cdots a'_{nn}$ prove the following: if $A = (a_{ij})$ is any symmetric and positive definite matrix, then $\det(A) \le a_{11} \cdots a_{nn}$. To show this, let D be the diagonal matrix whose diagonal is $1/\sqrt{a_{ii}}$. Consider DAD and once again using the arithmetic-geometric inequality prove that $\det(DAD) \le 1$.
To show equality assume A is diagonal and use Lagrange multipliers.

(5) Deduce from Problem 4 the following inequality due to Minkowski: if A, B are symmetric non negative definite $n \times n$ matrices, then

$$(\det(A + B))^{1/n} \ge (\det A)^{1/n} + (\det B)^{1/n};$$

that is, the function $\det A$ is concave over the non negative definite symmetric matrices. Equality happens iff A is a multiple of B.
HINT (by Ahmad Sabra): for the equality assume $\det A > 0$. Write $A = ODO^t$ with O orthogonal, and D diagonal with entries $\lambda_1, \cdots, \lambda_n$. Let $C = OD'O^t$ with D' diagonal with entries $\sqrt{\lambda_1}, \cdots, \sqrt{\lambda_n}$. Then C is symmetric and positive definite. Mutiply the identity by $\left(\det C^{-1}\right)^{1/n}$ to get $\left(\det\left(I + C^{-1}BC^{-1}\right)\right)^{1/n} = 1 + \left(\det\left(C^{-1}BC^{-1}\right)\right)^{1/n}$. Notice $C^{-1}BC^{-1}$ is symmetric, diagonalize it and conclude it is a multiple of the identity.

(6) Let A be an $n \times n$ matrix and $u \in C^2$. Let $v(x) = u(Ax)$. Prove that $D^2v(x) = A^t \left((D^2u)(Ax)\right) A$. Therefore $\det D^2v(x) = (\det A)^2 \det(D^2u)(Ax)$.

(7) Prove that $\partial u(x_0)$ is a convex set.

(8) Let $\Omega \subset \mathbb{R}^n$ be a closed convex set and let $y \notin \Omega$, and let $x_0 \in \Omega$ be the point that realizes the distance, i.e., $\text{dist}(y, \Omega) = |y - x_0|$. Prove that the hyperplane through x_0 with normal direction $y - x_0$ supports Ω at x_0.

(9) Let $\Omega \subset \mathbb{R}^n$ be closed and convex. Prove that for each x on the boundary of Ω there exists a supporting hyperplane to Ω at x. Proceed in steps:

   (a) prove first that if Ω is contained in the interior of a ball B, then for each $x \in \partial\Omega$ there exists $y \in \partial B$ such that $\text{dist}(y, \Omega) = |x - y|$.
   (b) from (a) obtain the result when Ω is bounded.
   (c) when Ω is unbounded, cut Ω with a ball and use (b).

   HINT: let x be on the boundary of K, with K convex closed and bounded, there is a ball B containing K in the interior. Given k, consider the ball $B(x, 1/k)$, since x is on the boundary of K, there is a point y_k in this ball that is not in K. Take distance $d(y_k, K)$. There is a unique point $x_k \in K$ such that $d(y_k, K) = |x_k - y_k|$, we have $|x_k - y_k| \le 2/k$.
   The point x_k is unique because K is convex, if there are two different points realizing the distance, the average point has distance from y_k strictly less than $d(y_k, K)$. Now we are going to find the point on the boundary of B. Take the ray emanating from x_k and passing through y_k. This ray intersects the boundary of B at some z_k. Notice that $d(z_k, K) = |z_k - x_k|$. Since the

boundary of B is compact, there is a subsequence z_k converging to some z in the boundary B. Then $d(z_k, K)$ converges to $d(z, K)$, and $|z_k - x_k| \to |z - x|$.

(10) Let $f : \Omega \to \mathbb{R}$ be a convex function. Prove that the subdifferential $\partial f(x) \neq \emptyset$ for each $x \in \Omega$.
HINT: Consider the set $\{(x, y) : x \in \Omega,\ y \geq f(x)\}$ and apply Problem 9.

(11) If f is convex in Ω and $\partial f(x_0)$ is a single point, then prove that f is differentiable at x_0.

Detailed Solution

Step 1 let us first prove that if a one variable convex function $h(t) \geq 0$, with $h(0) = 0$, is such that its subdifferential at $t = 0$ has only one point, i.e., $\partial h(0) = \{p\}$, then h is differentiable at $t = 0$. Suppose $0 < s < t$, then $0 < s/t < 1$ and since h is convex

$$h(s) = h\left(\frac{s}{t}t + \left(1 - \frac{s}{t}\right)0\right) \leq \frac{s}{t}h(t)$$

obtaining that

$$\frac{h(s)}{s} \leq \frac{h(t)}{t}, \qquad 0 < s < t.$$

Suppose now that $t < s < 0$, then $0 < s/t < 1$ so again by convexity

$$h(s) = h\left(\frac{s}{t}t + \left(1 - \frac{s}{t}\right)0\right) \leq \frac{s}{t}h(t)$$

and since $s < 0$ we get that

$$\frac{h(s)}{s} \geq \frac{h(t)}{t}, \qquad t < s < 0.$$

This means the function $h(s)/s$ is non decreasing in the intervals $(-\infty, 0)$, $(0, +\infty)$, $h(s)/s \geq 0$ in $(0, +\infty)$ and $h(s)/s \leq 0$ in $(-\infty, 0)$. So the following limits are finite

$$\lim_{s \to 0^+} \frac{h(s)}{s} = \inf_{s>0} \frac{h(s)}{s} = a \geq 0, \qquad \lim_{s \to 0^-} \frac{h(s)}{s} = \sup_{s<0} \frac{h(s)}{s} = b \leq 0,$$

so $b \leq 0 \leq a$. If we prove $a = b = 0$, then h is differentiable at $t = 0$ and $h'(0) = 0$. We have $h(s) \geq as$ for $s > 0$ and also $h(s) \geq as$ for $s < 0$ since $h \geq 0$ and $a \geq 0$. This means that $a \in \partial h(0)$ but since the only point in the subdifferential of h is p we get $a = p$. On the other hand, $h(s)/s \leq b$ for $s < 0$, so $h(s) \geq bs$ for $s < 0$ and $h(s) \geq bs$ for $s > 0$ since $h(s) \geq 0$ and $b \leq 0$. Therefore $b \in \partial h(0)$, and we again get $b = p$. So $a = b = p = 0$ and we are done.

Step 2 Now take f in $\mathbb{R}^n$ such $\partial f(x_0)$ is only one point p. Let $g(x) = f(x_0 + x) - f(x_0) - p \cdot x$. We have $g(x) \geq 0$ for all x, $g(0) = 0$, and g is convex. The subdifferential of $\partial g(0) = \{0\}$ because if $q \in \partial g(0)$ then $p + q \in \partial f(x_0)$ so $p + q = p$ and then $q = 0$.
Fix $x \neq 0$ and let $h(t) = g(tx)$ as a function of one variable t. We have $h \geq 0$, $h(0) = 0$, and h is convex. To apply Step 1 to h we need to show that $\partial h(0)$ is only one point. Clearly $0 \in \partial h(0)$. Let $b \in \partial h(0)$ and suppose $b \neq 0$. This means $g(t\,x) \geq b\,t$ for all $t \in \mathbb{R}$. Consider the line ℓ in $\mathbb{R}^{n+1}$ whose parametric equation is $(t\,x, b\,t)$, $t \in \mathbb{R}$. This line supports the graph of h at $t = 0$. Any plane Π through the origin has equation $z = p \cdot x$ for some p. If Π contains ℓ it means that $b\,t = p \cdot t\,x$, that is, $b = p \cdot x$. So if $b \neq 0$, then $p \neq 0$. Since the function g is convex in $\mathbb{R}^n$ and $g \geq 0$, there exists a plane Π containing ℓ that is a supporting plane to g at 0. Since $\partial g(0) = \{0\}$, we must have $p = 0$ so $b = 0$, a contradiction. Therefore $b = 0$. From Step 1, we get that h is differentiable at $t = 0$ and $h'(0) = 0$. This means that

$$\lim_{t \to 0} \frac{g(tx)}{t} = 0,$$

for each $x \neq 0$ fixed. Now the ball $B_1(0) \subset [-1, 1]^n$ and the diameter of $[-1, 1]^n$ equals $2\sqrt{n}$, so $[-1, 1]^n \subset B_{2^n}(0)$. Denote by v_i the vertices of $[-1, 1]^n$ and notice that $v_i \notin B_1(0)$. Since the convex hull of v_i equals $[-1, 1]^n$, each point $x \in B_1(0)$ can be written as a convex combination of v_i, that is, $x = \sum_{i=1}^m \lambda_i v_i$ with $\sum_{i=1}^m \lambda_i = 1$, $0 \leq \lambda_i \leq 1$ (notice that only λ_i depend on x). Then by convexity of g, and for $t > 0$ we have

$$\frac{g(tx)}{t} \leq \sum_{j=1}^m \lambda_i \frac{g(tv_i)}{t} \leq \sum_{j=1}^m \frac{g(tv_i)}{t},$$

with the right hand side of this inequality independent of x. Then $g(tx)/t$ tends to zero uniformly for x in the unit ball as $t \to 0^+$. This means that for each $\epsilon > 0$ there exists $\delta > 0$ such that

$$0 \leq \frac{g(tx)}{t} \leq \epsilon \tag{2.6}$$

for all $t \in (0, \delta)$ and for all $x \in \overline{B_1(0)}$. To show that g is differentiable at 0 we must show that

$$\lim_{y \to 0} \frac{g(y)}{|y|} = 0.$$

Write $g(y) = g\left(|y|\frac{y}{|y|}\right)$, and letting $x = \frac{y}{|y|} \in \overline{B_1(0)}$ from (2.6) we get that $g(|y|)/|y| \leq \epsilon$ as $|y| < \delta$ and we are done.

This implies f is differentiable at x_0.

(12) Suppose Ω is a bounded convex domain and u is bounded in Ω, so the Legendre transform u^* is finite in $\mathbb{R}^n$. Prove that

(a) $p_0 \in \partial u(x_0)$ iff $u(x_0) + u^*(p_0) = x_0 \cdot p_0$;
(b) if $p_0 \in \partial u(x_0)$, then $x_0 \in \partial u^*(p_0)$;
(c) we have $(u^*)^*(x) \leq u(x)$ for all $x \in \Omega$;
(d) if $\partial u(x_0) \neq \emptyset$, then $(u^*)^*(x_0) = u(x_0)$;
(e) if u is convex in Ω, then $p_0 \in \partial u(x_0)$ iff $x_0 \in \partial u^*(p_0)$;
(f) $\partial u(x)$ is a singleton for a.e. $x \in \Omega$ (by Lemma 2.7).
(g) if $u : \Omega \to \mathbb{R}$ is convex, then for each Borel set $F \subset \mathbb{R}^n$ the set

$$(\partial u)^{-1}(F) = \{x \in \Omega : \partial u(x) \cap F \neq \emptyset\}$$

is Lebesgue measurable. See Remark 5.2.
(h) if $u : \Omega \to \mathbb{R}$ is convex and differentiable in Ω, then

$$u^*(Du(x)) = x \cdot Du(x) - u(x)$$

for each $x \in \Omega$.
(i) if $\mathcal{N}u(x) = \{p \in \mathbb{R}^n : u(x) + u^*(p) = x \cdot p\}$, then $\mathcal{N}u(x) = \partial u(x)$.

(13) Suppose $u \in C^2(\Omega)$ is strictly convex in Ω. Let u^* be the Legendre transformation of u. Prove that $\det D^2u^*(p) = \dfrac{1}{\det D^2u(x)}$ with $p = Du(x)$.

(14) Prove that the subdifferential is a monotone map, that is, for all $x_1, x_2 \in \Omega$ and $p_i \in \partial u(x_i)$, $i = 1, 2$, we have $(x_1 - x_2) \cdot (p_1 - p_2) \geq 0$.

(15) Let T be a multivalued map defined for each $x \in \Omega$ such that $Tx \subset X$ where X is a fixed set. Prove that for each $E \subset \Omega$ we have

$$T(\Omega \setminus E) = (T(\Omega) \setminus T(E)) \cup (T(\Omega \setminus E) \cap T(E)).$$

(16) Let $T : \mathbb{R}^n \to \mathbb{R}^n$ be an inversible affine transformation, i.e., $Tx = Ax + b$ with A an $n \times n$ matrix, with $\det A \neq 0$, and $b \in \mathbb{R}^n$. Let $u : \Omega \to \mathbb{R}$, and set $v(x) = u(Tx)$ for $x \in T^{-1}\Omega$. Prove that

$$\partial v(x) = A^t\,(\partial u(Tx))$$

for all $x \in T^{-1}\Omega$.

(17) Let $z = u(x)$ be a convex hypersurface in $\mathbb{R}^{n+1}$. The Gauss map of u is given by

$$\mathcal{G}u(x) = \{\text{unit normals to each supporting hyperplane to } u \text{ at } (x, u(x))\}.$$

Prove that $p \in \partial u(x)$ if and only if $\dfrac{(p, -1)}{\sqrt{1 + |p|^2}} \in \mathcal{G}(x)$.

(18) Let $f \in C(\Omega)$ with Ω convex, and $R : \mathbb{R}^n \to \mathbb{R}$ continuous and both functions being positive. Suppose that $u \in C^2(\Omega) \cap C(\bar{\Omega})$ convex solves the problem $\det D^2u(x) = \dfrac{f(x)}{R(Du(x))}$. When $R(p) = (1 + |p|^2)^{(n+2)/2}$, we get the Gauss curvature equation.
Then prove that for each Borel set $E \subset \Omega$ we have

$$\int_E f(x)\,dx = \int_{\partial u(E)} R(p)\,dp,$$

and in particular, $\int_\Omega f(x)\,dx \le \int_{\mathbb{R}^n} R(p)\,dp$.
HINT: use Sard's Theorem 2.15 and the formula of change of variables.

(19) Let Ω be a bounded domain in $\mathbb{R}^n$ and μ is Borel measure over Ω such that $\mu(\Omega) < \infty$. Prove that

(a) for each k there exist a disjoint family $\{\Omega_j^k\}_{j=1}^{N_k}$ of Borel subsets of Ω, such that $\text{diam}(\Omega_j^k) < 1/k$ and $\Omega = \bigcup_{j=1}^{N_k} \Omega_j^k$.
(b) Pick $x_j^k \in \Omega_j^k$ and let $\mu_k = \sum_{j=1}^{N_k} \mu(\Omega_j^k)\,\delta_{x_j^k}$ with $\delta_{x_j^k}$ the Dirac delta concentrated at x_j^k. Prove that $\mu_k \to \mu$ weakly, that is, $\int_\Omega f\,d\mu_k \to \int_\Omega f\,d\mu$ as $k \to \infty$ for each $f \in C(\bar{\Omega})$.

(20) Let μ_n and μ be Borel measures in $\Omega \subset \mathbb{R}^n$ that are finite on compact sets (and therefore regular). Suppose that

(a) $\limsup_{k\to\infty} \mu_k(F) \le \mu(F)$ for each compact $F \subset \Omega$; and
(b) $\liminf_{k\to\infty} \mu_k(G) \ge \mu(G)$ for each open $G \subset \Omega$.

Prove that $\mu_k \to \mu$ weakly, that is, $\int_\Omega f(x)\,d\mu_k \to \int_\Omega f(x)\,d\mu$ for all f continuous with compact support in Ω (or for all f continuous and bounded in Ω if $\mu_k(\Omega)$ and $\mu(\Omega)$ are finite).
HINT: may assume $f[] \ge 0$. Given $\epsilon > 0$ choose numbers $\alpha_0 < \alpha_1 < \cdots < \alpha_N$ such that $\alpha_0 = 0$, $\sup_\Omega f(x) < \alpha_N$, and $\alpha_j - \alpha_{j-1} = \epsilon$ for $j = 1, \cdots, N$. First prove that

$$\limsup_{k\to\infty} \int_\Omega f(x)\,d\mu_k \le \int_\Omega f(x)\,d\mu.$$

Let $F_j = \{x \in \Omega : f(x) \ge \alpha_j\}$, for $j = 0, 1, \cdots, N$; and let $A_j = \{x \in \Omega : \alpha_{j-1} \le f(x) < \alpha_j\} = F_{j-1} \setminus F_j$, $j = 1, \cdots, N$; we have $F_j \subset F_{j-1}$ and $\sum_{j=1}^N \alpha_{j-1}\chi_{A_j}(x) \le f(x) \le \sum_{j=1}^N \alpha_j \chi_{A_j}(x)$. Integrate these inequalities with respect to μ_k, next with respect to μ, use (a), and compare the results. Second prove that

$$\liminf_{k\to\infty} \int_\Omega f(x)\,d\mu_k \ge \int_\Omega f(x)\,d\mu.$$

Let $G_j = \{x \in \Omega : f(x) > \alpha_j\}$ and $B_j = G_j \setminus G_{j-1}$; show that $\sum_{j=1}^{N} \alpha_{j-1}\chi_{B_j}(x) \leq f(x) \leq \sum_{j=1}^{N} \alpha_j \chi_{B_j}(x)$. Integrate these inequalities with respect to μ_k, next with respect to μ, use (b), and compare the results.

(21) Let E be an ellipsoid in $\mathbb{R}^n$ with center at the origin, and consider in $\mathbb{R}^{n+1}$ the upside down right cone with vertex at the origin with height h and top $E \times \{h\}$. Let v be the function from E to $\mathbb{R}$ whose graph is this cone. Calculate the subdifferential of v.

(22) Let U be convex in Ω, and $B \Subset \Omega$ be a ball. Prove that $\partial_\Omega U(E) = \partial_B U(E)$ for $E \subset B$, that is, the local and global subddiferential are equal on B (∂_B denotes the sub differential defined in B). This implies that the corresponding MA measures defined in Ω and B are equal in B.

(23) Let u, v be convex in Ω and $\phi = \max\{u, v\}$. If $\Omega' = \{x \in \Omega : u(x) < v(x)\}$, then $\partial\phi(E) = \partial v(E)$ for all $E \subset \Omega'$.

(24) Let $\mathcal{F}(\mu, g) = \{v \in C(\bar{\Omega}) : v \text{ convex}, Mv \geq \mu \text{ and } v = g \text{ on } \partial\Omega\}$ where Ω is a strictly convex domain, μ is a finite Borel measure in Ω and $g \in C(\partial\Omega)$. Prove that $\mathcal{F}(\mu, g)$ is a closed set of $C(\bar{\Omega})$ with the maximum norm.

Prove that $\mathcal{F}(\mu, g)$ is not compact in $C(\bar{\Omega})$.

HINT: By Arzela-Ascoli, a set has compact closure in $C(\bar{\Omega})$ if and only if is uniformly bounded and equicontinuous. Let $\Omega = B_1(0) \subset \mathbb{R}^n$, $\mu = \delta_0$. Consider the family of functions $v_h(x) = h\,(1 - |x|)$ with $h < 0$. Show that $Mv_h = \omega_n\,(-h)^n \delta_0$ where ω_n is the volume of the unit ball in $\mathbb{R}^n$. The family $\{v_h\}_{h \leq -\omega_n^{1/n}} \subset \mathcal{F}(\delta_0, 0)$ is not uniformly bounded.

(25) An ellipsoid in $\mathbb{R}^n$ with center at x_0 is a set of the form

$$E = \{x \in \mathbb{R}^n : \langle A(x - x_0), x - x_0 \rangle \leq 1\},$$

where A is an $n \times n$ positive definite symmetric matrix and $\langle\, , \rangle$ denotes the Euclidean inner product. Prove that

$$|E| = \frac{\omega_n}{(\det A)^{1/2}},$$

where ω_n is the volume of the unit ball.

Hint: formula of change of variables.

(26) If A, B are $n \times n$ symmetric positive definite matrices, then prove that

$$\det\left(\frac{A+B}{2}\right) \geq \sqrt{\det A \, \det B},$$

with equality iff $A = B$.

HINT: if O is an orthogonal matrix and $A' = O^t A O$, $B' = O^t B O$, then the inequality holds for A, B iff it holds for A', B'. So we may assume A is diagonal with diagonal $\lambda_1, \cdots, \lambda_n$. Let T be diagonal with diagonal $\sqrt{\lambda_1}, \cdots, \sqrt{\lambda_n}$. Show that the inequality holds for A, B iff it holds for $Id, T^{-1}BT^{-1}$. Then assume A is the identity and B is diagonal.

The inequality also follows from Minkowski's matrix inequality, Exercise 5, writing

$$\det\left(\frac{A+B}{2}\right) \geq \left((\det(A/2))^{1/n} + (\det(B/2))^{1/n}\right)^n$$
$$= \frac{1}{2^n}\left((\det A)^{1/n} + (\det B)^{1/n}\right)^n,$$

and using that $x^{1/n} + y^{1/n} \geq 2\sqrt{x^{1/n}y^{1/n}}$.

(27) Let K be an open bounded convex set in $\mathbb{R}^n$ (a convex body) with center of mass x_0. Consider all ellipsoids centered at x_0 and containing K, among these there exists an ellipsoid having minimum volume. Prove that this ellipsoid is unique.

HINT: suppose E_1 and E_2 are two different ellipsoids of minimum volume with corresponding defining matrices A_1 and A_2; we have $|E_1| = |E_2|$ and $A_1 \neq A_2$. Therefore, from Problem 25 $\det A_1 = \det A_2$. Consider the ellipsoid E with corresponding matrix $A = \dfrac{A_1 + A_2}{2}$. Prove that E contains K, and use Problems 25 and 26 to show that $|E| < |E_1|$.

(28) In $\mathbb{R}^3$ consider the function

$$u(x_1, x_2, x_3) = (1 + x_1^2)(x_2^2 + x_3^2)^{2/3}.$$

The objective is to show that u is a convex Aleksandrov solution to $Mu = \phi$ with $\phi \in C^\infty$, $\lambda \leq \phi \leq \Lambda$ (λ, Λ some positive constants) in a sufficiently small ball around the origin $B_\epsilon(0)$, u on $\partial B_\epsilon(0)$ is continuous, and $u \in C^1(B_\epsilon(0))$ but $u \notin C^2(B_\epsilon(0))$.

(a) calculate $D^2u(x_1, x_2, x_3)$ when $(x_1, x_2, x_3) \neq 0$;
(b) show that $\det D^2u(x_1, x_2, x_3) = 32/9\,(1 + x_1^2)\,((1/3) - (7/3)x_1^2) := \phi(x_1, x_2, x_3)$ for $((x_1, x_2, x_3) \neq 0$;
(c) the function u is not convex in a domain sufficiently far from zero.
(d) calculate the principal minors of D^2u when $(x_1, x_2, x_3) \neq 0$;
(e) prove that for $x_1^2 + x_2^2 + x_3^2 = \epsilon$ with ϵ sufficiently small, the determinants of the principal minors are positive and therefore $D^2u(x_1, x_2, x_3)$ is positive definite for all $(x_1, x_2, x_3) \neq 0$ in the ball $B_{\sqrt{\epsilon}}(0) = \{x_1^2 + x_2^2 + x_3^2 \leq \epsilon\}$.

HINT: for the principal minor of order two use Lagrange multipliers.
(f) show that the function u is strictly convex in any convex domain not intersecting the line $\ell \equiv \{x_2 = 0, x_3 = 0\}$ and contained in $B_{\sqrt{\epsilon}}(0)$, and it is C^∞ away from the line ℓ.
(g) the function u is convex in $B_{\sqrt{\epsilon}}(0)$. HINT: use that if a one variable function f is continuous non negative in $[-a, a]$, $f(0) = 0$, f is convex in $(0, a)$ and in $(-a, 0)$, then f is convex in $[-a, a]$. Use this as follows. If P_1, P_2 are in $B_{\sqrt{\epsilon}}(0)$, then we need to show the function u restricted to

the segment $\overline{P_1 P_2}$ is convex as a function of one variable. There are two cases: (a) when the segment $\overline{P_1 P_2}$ does not intersect ℓ, then the convexity follows from (f); if $\overline{P_1 P_2}$ intersects ℓ, then use the convexity result in one variable.

(h) the graph of u contains the line $x_2 = 0, x_3 = 0$;

(i) $\partial u(\{(x_1, 0, 0)\})$ has measure zero. HINT: use Aleksandrov's Lemma 2.7 (the set of supporting hyperplanes containing a segment in the graph of u has measure zero).

(j) if $E \subset B_{\sqrt{\epsilon}}(0)$ is a Borel set, then $|\partial u(E)| = \int_E \phi(x)\, dx$. HINT: $\partial u(E) = \partial u(E \cap \ell) \cup \partial u(E \cap \ell^c)$; and from (i) $|\partial u(E \cap \ell)| = 0$; next write E as a disjoint union over the eight octants and since u is C^2 and convex away from ℓ by adding over each piece of E we can represent $|\partial u(E \cap \ell^c)|$ as the integral of ϕ over $E \cap \ell^c$.

(k) $u \in C^{1,1/3}(B_{\sqrt{\epsilon}}(0))$ and $u \notin C^{1,\beta}(B_{\sqrt{\epsilon}}(0))$ for all $\beta > 1/3$.

Chapter 3
Sinkhorn's Theorem and Application to the Distribution Problem

Abstract The Sinkorn theorem is proved using Brouwer's fixed point theorem and the corresponding algorithm is then described and used to solve the distribution problem from Chap. 1. A connection with the Schrödinger bridge equations is presented.

The following lemma shows a way to obtain approximate values of the $\min_K x \cdot p$ by penalizing the objective function $x \cdot p$. In combination with Sinkhorn Theorem 3.2, this will be applied in Sect. 3.1 to solve the distribution problem stated in Sect. 1.1.

Lemma 3.1 *Let h be real valued strictly convex function in a region Ω of the real Hilbert space H with inner product $\langle , \rangle$, $\epsilon > 0$, $p \in H$, $K \subset \Omega$ compact and convex,*

$$F_\epsilon(x) = \langle x, p\rangle + \epsilon\, h(x),$$

$$E = \{x \in K : \langle x, p\rangle = \min_{y\in K}\langle y, p\rangle\},$$

and

$$\min_{x\in K} F_\epsilon(x) = F_\epsilon(x_\epsilon), \quad x_\epsilon \in K.$$

Let

$$m = \min_{x\in E} h(x).$$

Then there exists a unique $x_0 \in E$ with $m = h(x_0)$ and $x_\epsilon \to x_0$ as $\epsilon \to 0$.

Proof F_ϵ is strictly convex. Since K is convex, E is convex; and since K is compact, E is compact. Since h is strictly convex, there exists a unique $x_0 \in E$ such that $m = h(x_0)$. Indeed, if $x_1, x_2 \in E$, $x_1 \neq x_2$, with $h(x_1) = h(x_2) = m$, then by

C. E. Gutiérrez, *Optimal Transport and Applications to Geometric Optics*,
SpringerBriefs on PDEs and Data Science,
https://doi.org/10.1007/978-981-99-4867-3_3

the strict convexity $h\,(t\,x_1 + (1-t)\,x_2) < t\,h(x_1) + (1-t)\,h(x_2) = m$. Since E is convex $t\,x_1 + (1-t)\,x_2 \in E$ obtaining a contradiction.

Since K is compact, each subsequence of x_ϵ contains a subsequence converging to some point $y \in K$, a subsequence that will be also denoted by x_ϵ. We have

$$\langle x_\epsilon, p\rangle + \epsilon\, h(x_\epsilon) \le \langle x, p\rangle + \epsilon\, h(x) \qquad \forall x \in K. \tag{3.1}$$

For $x \in E$, $\langle x, p\rangle \le \langle x_\epsilon, p\rangle$, so

$$0 \le \langle x_\epsilon, p\rangle - \langle x, p\rangle \le \epsilon\,(h(x) - h(x_\epsilon)) \qquad \forall x \in E. \tag{3.2}$$

Hence $h(x) \ge h(x_\epsilon)$ for all $x \in E$, and letting $\epsilon \to 0$, by continuity $h(x) \ge h(y)$ for all $x \in E$ and so $m \ge h(y)$. Letting $\epsilon \to 0$ in (3.1) yields $\langle y, p\rangle \le \langle x, p\rangle$ for all $x \in K$, so $y \in E$ and $m = h(y)$, that is, $y = x_0$. □

Theorem 3.2 ([41, 53]) *Let A be an $m \times n$ real matrix with strictly positive entries and let $\mathbf{p} = (p_1, \cdots, p_m) \in \mathbb{R}^m$ and $\mathbf{q} = (q_1, \cdots, q_n) \in \mathbb{R}^n$ be non zero row vectors with non negative components satisfying*

$$\sum_{k=1}^{m} p_k = \sum_{\ell=1}^{n} q_\ell. \tag{3.3}$$

Then there exist unique[1] *diagonal matrices $D_1 \in \mathbb{R}^{m\times m}$, $D_2 \in \mathbb{R}^{n\times n}$ such that*

$$D_1 A D_2 \in \mathcal{N}(\mathbf{p}, \mathbf{q}).$$

Proof Fix $x = (x_1, \cdots, x_n) \neq 0$ with $x_i \ge 0$, consider the system of equations in $u = (u_1, \cdots, u_m)$

$$\begin{pmatrix} u_1 & 0 & \cdots & 0 \\ 0 & u_2 & \cdots & 0 \\ \vdots & \vdots & \ddots & \vdots \\ 0 & 0 & \cdots & u_m \end{pmatrix} A(x_1, \cdots, x_n)^t = (p_1, \cdots, p_m)^t, \tag{3.4}$$

and with that value of u consider the system of equations in $y = (y_1, \cdots, y_n)$

$$\begin{pmatrix} y_1 & 0 & \cdots & 0 \\ 0 & y_2 & \cdots & 0 \\ \vdots & \vdots & \ddots & \vdots \\ 0 & 0 & \cdots & y_n \end{pmatrix} A^t(u_1, \cdots, u_m)^t = (q_1, \cdots, q_n)^t. \tag{3.5}$$

[1] Uniqueness is up to a multiple, i.e., μD_1 and $\mu^{-1} D_2$ also verify the theorem.

So (3.4) reads

$$u_k \sum_{j=1}^{n} a_{kj}\, x_j = p_k, \quad 1 \le k \le m, \tag{3.6}$$

which yields

$$u_k = \frac{p_k}{\sum_{j=1}^{n} a_{kj}\, x_j}, \quad 1 \le k \le m.$$

Notice that since all the $a_{kj} > 0$ and $x \neq 0$, $x_i \geq 0$, the denominator of the last fraction is strictly positive so $u_k \geq 0$. In addition, since $\mathbf{p} \neq 0$, the vector u is not zero. Also (3.5) reads

$$y_k \sum_{\ell=1}^{m} a_{\ell k}\, u_\ell = q_k, \quad 1 \le k \le n \tag{3.7}$$

which yields

$$y_k = \frac{q_k}{\sum_{\ell=1}^{m} a_{\ell k}\, u_\ell}, \quad 1 \le k \le n.$$

Once again since the $a_{kj} > 0$ and $u \neq 0$, the last denominator is not zero and since $\mathbf{q} \neq 0$, we get that $y \neq 0$. So we have a mapping $T : \mathbb{R}^n_+ \to \mathbb{R}^n_+$ defined by $Tx = y$ with

$$y_k = \frac{q_k}{\sum_{\ell=1}^{m} a_{\ell k} \dfrac{p_\ell}{\sum_{j=1}^{n} a_{\ell j} x_j}}, \quad 1 \le k \le n.$$

If $\lambda > 0$, then $T(\lambda x) = \lambda Tx$. Consider the part of the unit sphere in the first quadrant

$$E = \{x = (x_1, \cdots, x_n) : |x| = 1, x_i \geq 0\}.$$

E is a compact set homeomorphic to a compact convex set in R^{n-1}. Define

$$F(x) = \frac{Tx}{|Tx|}.$$

We have that $F : E \to E$, and F is continuous in E. Then by Brouwer's fixed point theorem, there exists $r \in E$ with $F(r) = r$. We claim that $Tr = r$. In fact, we first have $Tr = |Tr|\, r$. Let us show that $|Tr| = 1$. Set $y = |Tr| r$. From (3.7)

$$\sum_{k=1}^{n} y_k \sum_{\ell=1}^{m} a_{\ell k}\, u_\ell = \sum_{k=1}^{n} q_k,$$

so

$$\begin{aligned}\sum_{k=1}^{n} q_k &= |Tr| \sum_{k=1}^{n} r_k \sum_{\ell=1}^{m} a_{\ell k}\, u_\ell \\ &= |Tr| \sum_{\ell=1}^{m} u_\ell \sum_{k=1}^{n} a_{\ell k}\, r_k \\ &= |Tr| \sum_{\ell=1}^{m} p_\ell\end{aligned}$$

from (3.6) and so the claim follows from (3.3).

Thus taking $x = r$ in (3.4) we get the corresponding solution u, which inserted in (3.5) yields the solution $y = r$. If we let D_1 to be the $m \times m$ matrix with diagonal u and D_2 to be the $n \times n$ matrix with diagonal r we obtain that $D_1 A D_2 \in \mathcal{N}(\mathbf{p}, \mathbf{q})$.

The proof of uniqueness is in [53]. □

3.1 Application to the Distribution Problem

We now apply Lemma 3.1 to the functional

$$F_\epsilon(A) = \langle A, C\rangle + \epsilon\, h(A)$$

with the compact $K = \mathcal{N}(\mathbf{u}, \mathbf{v})$ described with the constraints on A given by (1.1), and where h is a convex function in the region $a_{ij} \ge 0$; $A = (a_{ij})$. We then introduce the Lagrangian with the multipliers $\lambda = (\lambda_1, \cdots, \lambda_m)$ and $\mu = (\mu_1, \cdots, \mu_n)$, that is,

$$L(A, \lambda, \mu) = \langle A, C\rangle + \epsilon\, h(A) + \sum_{i=1}^{m} \lambda_i \left(\sum_{j=1}^{n} a_{ij} - u_i \right) + \sum_{j=1}^{n} \mu_j \left(\sum_{i=1}^{m} a_{ij} - v_j \right).$$

Suppose $\min_{A \in \mathcal{N}(\mathbf{u},\mathbf{v})} F_\epsilon(A)$ is attained at A_ϵ. Then from Lagrange multipliers theorem, there exist λ^0 and μ^0 such that

$$\frac{\partial L}{\partial a_{ij}} \left(A_\epsilon, \lambda^0, \mu^0 \right) = 0,$$

that is,

$$\frac{\partial L}{\partial a_{ij}} \left(A_\epsilon, \lambda^0, \mu^0 \right) = c_{ij} + \epsilon\, \frac{\partial h}{\partial a_{ij}} (A_\epsilon) + \lambda_i + \mu_j = 0$$

obtaining that

$$\frac{\partial h}{\partial a_{ij}}(A_\epsilon) = -\frac{1}{\epsilon}\left(c_{ij} + \lambda_i + \mu_j\right) := B_\epsilon.$$

Let h^* be the Legendre transform of h. Since h is strictly convex, $P = \nabla h(A)$ if and only if $A = \nabla h^*(P)$. So the point A_ϵ where the minimum is attained satisfies

$$(A_\epsilon)_{ij} = \frac{\partial h^*}{\partial p_{ij}}(B_\epsilon). \tag{3.8}$$

Let us consider the special case when $h(A) = \sum_{ij} a_{ij}\left(\log a_{ij} - 1\right)$; this is a strictly convex function when $a_{ij} > 0$. By definition

$$h^*(P) = \sup_A \{\langle A, P\rangle - h(A)\} = \sup_A \left\{\sum_{ij} a_{ij}\, p_{ij} - \sum_{ij} a_{ij}\left(\log a_{ij} - 1\right)\right\}.$$

Now at the maximum $\partial_{a_{ij}}\left(\sum_{ij} a_{ij}\, p_{ij} - \sum_{ij} a_{ij}\left(\log a_{ij} - 1\right)\right) = 0$ which gives $p_{ij} - \log a_{ij} = 0$, that is, $a_{ij} = e^{p_{ij}}$. Therefore

$$h^*(P) = \sum_{ij} e^{p_{ij}}$$

and so

$$\frac{\partial h^*}{\partial p_{ij}} = e^{p_{ij}}.$$

Thus from (3.8) the point $A_\epsilon \in \mathcal{N}(\mathbf{u}, \mathbf{v})$ (a set defined after Eq. (1.1)), where the minimum of F_ϵ is attained, satisfies

$$(A_\epsilon)_{ij} = \exp\left(-\frac{1}{\epsilon}\left(c_{ij} + \lambda_i + \mu_j\right)\right) = \exp\left(-\lambda_i/\epsilon\right)\exp\left(-c_{ij}/\epsilon\right)\exp\left(-\mu_i/\epsilon\right). \tag{3.9}$$

To find A_ϵ we only need to find the diagonal matrices $\exp(-\lambda_i/\epsilon)$ and $\exp(-\mu_i/\epsilon)$. For this we apply Sinkhorn's Theorem 3.2 with the matrix $A = \exp\left(-c_{ij}/\epsilon\right)$. Indeed, since $A_\epsilon \in \mathcal{N}(\mathbf{u}, \mathbf{v})$, (3.9) implies that the diagonal matrices $\exp(-\lambda_i/\epsilon)$ and $\exp(-\mu_i/\epsilon)$ are uniquely determined, up to a multiple, by the matrix $\exp\left(-c_{ij}/\epsilon\right)$. So if we have an effective way to find the matrices D_1 and D_2 in Sinkhorn's Theorem 3.2, then we would obtain A_ϵ, see Remark 3.2 below.

3.2 Sinkhorn's Algorithm

From the idea of proof of Theorem 3.2 we obtain the following iterative method to find the required matrices D_1 and D_2. Given a vector $z = (z_1, \cdots, z_n)$, $D(z)$ denotes the $n \times n$ matrix having diagonal z and zeros otherwise. So Eqs. (3.4) and (3.5) can be written as

$$D(u)Ax^t = p^t \tag{3.10}$$

$$D(y)A^t u^t = q^t. \tag{3.11}$$

Given $x \in \mathbb{R}^n_+$ denote $u = G(x)$ be the solution to (3.10), so $G : \mathbb{R}^n_+ \to \mathbb{R}^m_+$. And given $u \in \mathbb{R}^m_+$, let $y = H(u)$ be the solution to (3.11), so $H : \mathbb{R}^m_+ \to \mathbb{R}^n_+$. Notice that in the proof of Theorem 3.2, the mapping $T = HG$. Then for the iterative process, fix an initializing vector $x = x_1 \in \mathbb{R}^n_+$ and let $u_1 = G(x_1)$ be the solution to (3.10). Insert $u = u_1$ in (3.11), and let $y_1 = Hu_1 = HG(x_1)$ be the solution to (3.11). Next let $x = y_1 = HG(x_1)$ in (3.10), and let u_2 solve (3.10). That is, $u_2 = Gy_1 = GHG(x_1)$. Inserting $u = u_2$ in (3.11) and solving in y yields $y = y_2 = Hu_2 = HGHG(x_1)$. Continuing in this way, we get sequences $u_k = G(HG)^{k-1}(x_1)$ and $y_k = Hu_k = HG(HG)^{k-1}(x_1) = (HG)^k(x_1)$. It can be proved that these sequences converge as $k \to \infty$ to values $\bar{u}$ and $\bar{y}$, respectively, see [46, pp. 63–71] for a discussion of the various approaches to prove this convergence and applications.[2] Taking the convergence for granted, from the definitions of u_k and y_k we have

$$D(u_k)A\,(y_{k-1})^t = p^t$$

$$D\,(y_k)\,A^t\,(u_k)^t = q^t,$$

and letting $k \to \infty$ in these equations yield

$$D(\bar{u})A\,(\bar{y})^t = p^t$$

$$D\,(\bar{y})\,A^t\,(\bar{u})^t = q^t.$$

This means $\bar{u} = G\bar{y}$ and $\bar{y} = H\bar{u}$, so $\bar{y} = HG\bar{y}$, that is, $\bar{y}$ is a fixed point of the map $T = HG$. Therefore as at the end of the proof of Theorem 3.2, the diagonal matrices $D_1 = D(\bar{u})$ and $D_2 = D(\bar{y})$ do the desired job.

Remark 3.3 Let us consider $C(X)$ with the maximum norm, where X is compact, μ is a finite Borel measure in X, and let $h : X \times X \to \mathbb{R}^+$ be a Borel measurable function satisfying

$$0 < A \leq h(x, y) \leq B$$

[2] For a version of this problem and its convergence in infinite dimensions see [47].

for all $(x, y) \in X \times X$ with A, B constants. Given $f \in C(X)$ define the operator

$$Tf(x) = \int_X h(x, y)\, f(y)\, d\mu(y).$$

Also let $p, q \in C(X)$ be given, $p, q \geq 0$, not both identically zero. Consider the problem of finding functions $f, g, k \in C(X)$ non negative and not identically zero such that

$$g(x)\, Tf(x) = p(x) \tag{3.12}$$

$$k(x)\, Tg(x) = q(x). \tag{3.13}$$

These are a continuous version of Eqs. (3.4) and (3.5). Define the mappings

$$Pf(x) = \frac{p(x)}{Tf(x)} \qquad Qf(x) = \frac{q(x)}{Tf(x)}.$$

So if g solves (3.12), then $g(x) = Pf(x)$ and if k solves (3.13), then $k(x) = Qg(x)$. Therefore $k(x) = QPf(x)$. Notice that if $f \geq 0$ and $f \neq 0$, then

$$A\, \|f\|_{L^1_\mu} \leq Tf(x) \leq B\, \|f\|_{L^1_\mu} \quad \forall x \in X.$$

So if $\|f\|_{L^1_\mu} \neq 0$, then

$$\frac{p(x)}{B\, \|f\|_{L^1_\mu}} \leq Pf(x) \leq \frac{p(x)}{A\, \|f\|_{L^1_\mu}}$$

and

$$\frac{q(x)}{B\, \|f\|_{L^1_\mu}} \leq Qf(x) \leq \frac{q(x)}{A\, \|f\|_{L^1_\mu}}.$$

Therefore

$$\left(\frac{A}{B}\right) \frac{\|f\|_{L^1_\mu}}{\|p\|_{L^1_\mu}}\, q(x) \leq QPf(x) = \frac{q(x)}{T\,(Pf)\,(x)} \leq \frac{q(x)}{A\, \|Pf\|_{L^1_\mu}}$$

$$\leq \left(\frac{B}{A}\right) \frac{\|f\|_{L^1_\mu}}{\|p\|_{L^1_\mu}}\, q(x).$$

Now, if we can find f, g, k with $f = k$ solving (3.12) and (3.13), then $QPf = f$. Vice versa, if there is f satisfying $QPf = f$, then taking $g = Pf$ and $k = Qg$, we have that f, g, k solve (3.12) and (3.13). Therefore, the problem of finding f, g, k

with $f = k$ solving (3.12) and (3.13) equivalent to find a fixed point of the mapping QP. We can also write

$$QPf(x) = \frac{q(x)}{TPf(x)} = \frac{q(x)}{\int h(x,y)Pf(y)\,d\mu(y)}$$
$$= \frac{q(x)}{\int h(x,y)\dfrac{p(y)}{\int h(y,z)f(z)\,d\mu(z)}\,d\mu(y)}.$$

Since p, q, h are all nonnegative, $QPf_1 \le QPf_2$ if $f_1 \le f_2$. Also, if the functions p, q, h are continuous and p, q are both strictly positive, then if $f_1 \le f_2$ are continuous and $f_1 \neq f_2$, we have $QPf_1(x) \lneqq QPf_2(x)$ for all $x \in X$. Equations (3.12) and (3.13) are the so called Schrödinger bridge equations arising in a problem concerning Brownian motion [51, Sec. VII, Eq. (9)]. Existence of solutions was proved by Fortet [18] using a successive approximation method, and in a more general setting by Beurling [3]. See the beautiful recent paper [16, Sect. 4, Eq. (S)] for a clear presentation of Fortet's method, connections with other problems, and extended references.

Chapter 4
Monge-Kantorovich Distance

Abstract This Chapter begins with the notion of disintegration of measures and states the disintegration theorem fully proved later in Chap. 14. The Monge-Kantorovich or Wasserstein distance is introduced and its properties proved using disintegration.

4.1 Disintegration of Measures

The notion of disintegration of measures goes back to von Neumann [56]. See [6] for a high level complete presentation, the paper [11] for a nice presentation showing ideas and ramifications of the notion of disintegration together with historical remarks, and [13, Chap. III, p. 78].

Definition 4.1 (Measurable Map) Let X, Y be locally compact spaces with a countable basis, and μ a Borel measure in X. The function $T : X \to Y$ is a measurable map if $T^{-1}(E)$ is a Borel set of X for each E Borel set of Y.

Definition 4.2 (Disintegration Family) A family $\{\lambda_y\}_{y\in Y}$ of Borel measures in X is a disintegration family of the measure μ relative to the measurable map $T : X \to Y$ if it satisfies the following properties:

(1) the function $y \mapsto \lambda_y(B)$ is Borel measurable in Y for each Borel set $B \subset X$
(2) $\lambda_y\left(X \setminus T^{-1}(y)\right) = 0$ for each $y \in Y$, i.e., λ_y is concentrated on the set $T^{-1}(y)$
(3) $\mu = \int \lambda_y\, d\nu(y)$ for $\nu = T_{\#}\mu$, i.e., $\mu(E) = \int \lambda_y(E)\, d\nu(y)$ for each $E \subset X$ Borel measurable.

Theorem 4.3 ([6, §3, no. 1, Th. 1, pp. 58-59]) *Let X, Y be locally compact spaces with a countable basis, μ a Borel measure in X, let $T : X \to Y$ be a measurable map and $\nu = T_{\#}\mu$.*

Then there exists a disintegration family $\{\lambda_y\}_{y\in Y}$ of the measure μ relative to the map T and this family is unique a.e. in ν, i.e., if $\{\lambda'_y\}_{y\in Y}$ is another disintegration family then $\lambda'_y = \lambda_y$ for a.e. $y \in Y$ in the measure ν.

C. E. Gutiérrez, *Optimal Transport and Applications to Geometric Optics*,
SpringerBriefs on PDEs and Data Science,
https://doi.org/10.1007/978-981-99-4867-3_4

As a consequence of this theorem we show the following lemma which will be used to prove that the Monge-Kantorovich distance given in Definition 4.6 satisfies the triangle inequality, Theorem 4.7. Theorem 4.3 follows from Theorem 14.11.

Lemma 4.4 (Gluing) *Let μ_i be Borel probability measures in X, $1 \leq i \leq 3$. Let $\gamma^{12} \in \Pi(\mu_1, \mu_2)$, $\gamma^{23} \in \Pi(\mu_2, \mu_3)$. Denote by $\pi^{ij} : X \times X \times X \to X \times X$ the projections, i.e., $\pi^{12}(x_1, x_2, x_3) = (x_1, x_2)$ and $\pi^{23}(x_1, x_2, x_3) = (x_2, x_3)$. There exists a probability measure γ in $X \times X \times X$ such that*

$$(\pi^{12})_{\#}\gamma = \gamma^{12}, \quad (\pi^{23})_{\#}\gamma = \gamma^{23}. \tag{4.1}$$

Proof Consider first the measure γ^{12} in $X \times X$ and disintegrate this measure relative to the projection map $\pi^2 : X \times X \to X$ given by $\pi^2(x, y) = y$. From Theorem 4.3 we then get a family of probability measures $\{\lambda_y^{12}\}_{y \in X}$ satisfying

$$\gamma^{12} = \int_X \lambda_y^{12}\, d\nu(y), \quad \text{with } \nu = (\pi^2)_{\#}\gamma^{12} = \mu_2, \tag{4.2}$$

since $\gamma^{12} \in \Pi(\mu_1, \mu_2)$. In addition, since $(\pi^2)^{-1}(y) = X \times \{y\}$ and λ_y^{12} is concentrated there, we have $\lambda_y^{12}(E) = \lambda_y^{12}(E \cap (X \times \{y\}))$ for each measurable set $E \subset X \times X$.

Second consider the measure γ^{23} in $X \times X$ and disintegrate this measure relative to the map $\pi^1 : X \times X \to X$ given by $\pi^1(x, y) = x$. Again from Theorem 4.3 we get a family of probability measures $\{\lambda_y^{23}\}_{y \in X}$ in $X \times X$ satisfying

$$\gamma^{23} = \int_X \lambda_y^{23}\, d\nu(y), \quad \text{with } \nu = (\pi^1)_{\#}\gamma^{23} = \mu_2,$$

since $\gamma^{23} \in \Pi(\mu_2, \mu_3)$. Also, since now $(\pi^1)^{-1}(y) = \{y\} \times X$ and λ_y^{23} is concentrated there, we get $\lambda_y^{23}(E) = \lambda_y^{23}(E \cap (\{y\} \times X))$ for each measurable set $E \subset X \times X$.

Now define the following measures in X as follows

$$\overline{\lambda_y^{12}}(E) = \lambda_y^{12}(E \times \{y\}),$$
$$\overline{\lambda_y^{23}}(E) = \lambda_y^{23}(\{y\} \times E), \quad \text{for each } E \subset X \text{ measurable.}$$

And next for each $y \in X$ take the product measure[1] $\overline{\lambda_y^{12}} \otimes \overline{\lambda_y^{23}}$ in $X \times X$ and define the measure γ in $X \times X \times X$ as follows

$$\gamma(A \times B \times C) = \int_B \left(\overline{\lambda_y^{12}} \otimes \overline{\lambda_y^{23}}\right)(A \times C)\, d\mu_2(y).$$

---

[1] $(\mu \otimes \nu)(A \times B) = \mu(A)\,\nu(B)$.

We will show that γ is the desired measure satisfying (4.1). Notice that from the definitions we have for $E \subset X \times X$

$$(\pi^{12})_{\#}\gamma(E) = \gamma\left((\pi^{12})^{-1}(E)\right) = \gamma\,(E \times X)\,.$$

If $E = A \times B$, then by definition of γ

$$\begin{aligned}
\gamma\,(E \times X) &= \gamma\,(A \times B \times X)\\
&= \int_B \left(\overline{\lambda_y^{12}} \otimes \overline{\lambda_y^{23}}\right)(A \times X)\, d\mu_2(y)\\
&= \int_B \overline{\lambda_y^{12}}(A)\, \overline{\lambda_y^{23}}(X)\, d\mu_2(y)\\
&= \int_B \lambda_y^{12}\,(A \times \{y\})\; \lambda_y^{23}\,(\{y\} \times X)\; d\mu_2(y)\\
&= \int_B \lambda_y^{12}\,(A \times \{y\})\; d\mu_2(y) \quad \text{since } 1 = \lambda_y^{23}(X \times X) = \lambda_y^{23}\,(\{y\} \times X)\,.
\end{aligned}$$

On the other hand, from (4.2)

$$\begin{aligned}
\gamma^{12}(E) = \gamma^{12}(A \times B) &= \int_X \lambda_y^{12}(A \times B)\, d\mu_2(y)\\
&= \int_X \lambda_y^{12}\,((A \times B) \cap (X \times \{y\})\; d\mu_2(y)\\
&= \int_X \lambda_y^{12}\,((A \cap X) \times (B \cap \{y\}))\; d\mu_2(y)\\
&= \int_X \lambda_y^{12}\,(A \times (B \cap \{y\}))\; d\mu_2(y)\\
&= \int_B \lambda_y^{12}\,(A \times (B \cap \{y\}))\; d\mu_2(y)\\
&\quad + \int_{X \setminus B} \lambda_y^{12}\,(A \times (B \cap \{y\}))\; d\mu_2(y)\\
&= \int_B \lambda_y^{12}\,(A \times \{y\})\; d\mu_2(y),
\end{aligned}$$

so the first identity in (4.1) follows. The proof of the second identity in (4.1) is similar. □

Remark 4.5 We recall that we use here all the time the following simple but important fact: if $T : X \to Y$ is a measurable mapping and μ is a Borel measure on X, then set $T_{\#}\mu(E) = \mu\left(T^{-1}(E)\right)$ for each $E \subset Y$ Borel set. If we denote

$\sigma = T_{\#}\mu$, then we obviously have $\sigma(E) = \mu\left(T^{-1}(E)\right)$ for $E \subset Y$ which can be written as

$$\int_Y \chi_E(y)\, d\sigma(y) = \int_X \chi_{T^{-1}(E)}(x)\, d\mu(x) = \int_X \chi_E(Tx)\, d\mu(x)$$

since $\chi_{T^{-1}(E)}(x) = \chi_E(Tx)$. By linearity

$$\int_Y \sum_i \alpha_i\, \chi_{E_i}(y)\, d\sigma(y) = \int_X \sum_i \alpha_i \chi_{E_i}(Tx)\, d\mu(x)$$

and by the density of simple functions in $C(Y)$ we get

$$\int_Y \phi(y)\, d\sigma(y) = \int_X \phi(Tx)\, d\mu(x) \tag{4.3}$$

for each $\phi \in C(Y)$.

4.2 Wasserstein Distance

Definition 4.6 (Monge-Kantorovich Distance) Given μ, ν probability Borel measures in X, $1 \leq p < \infty$, and $d : X \times X \to [0, +\infty)$ a distance, the Monge-Kantorovich distance -also called Wasserstein distance- is defined by

$$W_p(\mu, \nu) = \min \left\{ \left(\int_{X \times X} d(x, y)^p\, d\gamma \right)^{1/p} : \gamma \in \Pi(\mu, \nu) \right\}.$$

We will show that the quantity defined is a metric on the set of probability measures over X. The difficult part is to show that W_p satisfies the triangle inequality.

Theorem 4.7 *Suppose d is a quasi distance in X, i.e., is symmetric, $d(x, y) = 0$ if and only if $x = y$, and d satisfies the triangle inequality with a constant K. Then quantity $W_p(\mu, \nu)$ is a quasi metric with constant K in the space of probability measures over the compact metric space X; provided d is continuous with respect to the topology in X.*

Proof

1. $W_p(\mu, \mu) = 0$. It follows from (1.3) taking $c(x, y) = d(x, y)^p$, because if $T : X \to X, Tx = x$, then $T \in \mathcal{S}(\mu, \mu)$ since $d(x, x) = 0$. [2]
2. *if* $W_p(\mu, \nu) = 0$, *then* $\mu = \nu$. If $W_p(\mu, \nu) = 0$, then from Theorem 1.4, there exists $\gamma \in \Pi(\mu, \nu)$ such that $\int_{X\times X} d(x, y)^p \, d\gamma(x, y) = 0$. Since d is continuous and $d(x, y) > 0$ for $x \neq y$, the support of the measure γ is contained in the set $\{(x, x) : x \in X\}$. Since $(E\times X)\cap\{(x, x) : x \in X\} = (X\times E)\cap\{(x, x) : x \in X\}$, and $\gamma \in \Pi(\mu, \nu)$, we get

$$\begin{aligned}\mu(E) &= \gamma(E \times X) = \gamma\left((E \times X) \cap \{(x, x) : x \in X\}\right)\\ &= \gamma\left((X \times E) \cap \{(x, x) : x \in X\}\right) = \gamma(X \times E) = \nu(E).\end{aligned}$$

3. $W_p(\mu, \nu) = W_p(\nu, \mu)$. Let $\gamma \in \Pi(\mu, \nu)$, $S : X \times X \to X \times X$ with $S(x, y) = (y, x)$, and consider $S_\#\gamma$. We have $S_\#\gamma(A \times B) = \gamma(S^{-1}(A \times B)) = \gamma(B \times A)$ so $\gamma \in \Pi(\mu, \nu)$ if and only if $S_\#\gamma \in \Pi(\nu, \mu)$. For each $f \in C(X \times X)$

$$\begin{aligned}\int_{X\times X} f(x, y)\, d(S_\#\gamma(x, y)) &= \int_{X\times X} f(S(x, y))\, d\gamma(x, y)\\ &= \int_{X\times X} f(y, x)\, d\gamma(x, y)\end{aligned}$$

and taking $f(x, y) = d(x, y)^p$, since $f(x, y) = f(y, x)$, we are done.
4. *Triangle inequality.* The proof uses disintegration of measures. From Theorem 1.4, there exist $\gamma^{13} \in \Pi(\mu_1, \mu_3)$ and $\gamma^{32} \in \Pi(\mu_3, \mu_2)$, both probability measures in $X \times X$, such that

$$W_p(\mu_1, \mu_3) = \left(\int_{X\times X} d(x, y)^p \, d\gamma^{13}\right)^{1/p}$$

and

$$W_p(\mu_3, \mu_2) = \left(\int_{X\times X} d(x, y)^p \, d\gamma^{32}\right)^{1/p}.$$

Applying Lemma 4.4 to the measures γ^{13} and γ^{32}, there exists a probability measure γ over $X \times X \times X$ such that

$$(\pi^{12})_\#\gamma = \gamma^{13}, \quad (\pi^{23})_\#\gamma = \gamma^{32}.$$

---

[2] This can be proved directly without appealing to the Monge problem. Indeed, let $S : X \to X \times X$ be defined by $Sx = (x, x)$ and let $\gamma(E) = S_\#\mu(E)$. Since $S^{-1}(A \times B) = A \cap B$, $\gamma(X \times B) = \mu(X \cap B) = \mu(B)$ and $\gamma(A \times X) = \mu(A \cap X) = \mu(A)$, so $\gamma \in \Pi(\mu, \mu)$. And $\int_{X\times X} d(x, y)^p \, d\gamma = \int_X d(Sx)^p \, d\mu = \int_X d(x, x)^p \, d\mu = 0$.

Let $\pi^{13} : X \times X \times X \to X \times X$ be the projection $\pi^{13}(x_1, x_2, x_3) = (x_1, x_3)$ and define $\sigma = \left(\pi^{13}\right)_{\#} \gamma$. So σ is a measure in $X \times X$. We claim that $\sigma \in \Pi(\mu_1, \mu_2)$, that is, we will show that $\sigma(A \times X) = \mu_1(A)$ and $\sigma(X \times B) = \mu_2(B)$ for all Borel sets $A, B \subset X$. Indeed,

$$\sigma(A \times X) = \left(\left(\pi^{13}\right)_{\#} \gamma\right)(A \times X) = \gamma\left(\left(\pi^{13}\right)^{-1}(A \times X)\right) = \gamma(A \times X \times X).$$

On the other hand

$$\left(\left(\pi^{12}\right)_{\#} \gamma\right)(A \times X) = \gamma\left(\left(\pi^{12}\right)^{-1}(A \times X)\right) = \gamma\,(A \times X \times X).$$

But $\left((\pi^{12})_{\#} \gamma\right)(A \times X) = \gamma^{13}(A \times X) = \mu_1(A)$ since $\gamma^{13} \in \Pi(\mu_1, \mu_3)$. Therefore $\sigma(A \times X) = \mu_1(A)$. Similarly, $\sigma(X \times B) = \gamma(X \times X \times B) = (\pi^{23})_{\#}\gamma(X \times B) = \gamma^{32}(X \times B) = \mu_2(B)$, since $\gamma^{32} \in \Pi(\mu_3, \mu_2)$. The claim is then proved.

Hence

$$W_p(\mu_1, \mu_2) \le \left(\int_{X \times X} d(x, y)^p \, d\sigma\right)^{1/p}.$$

Applying now (4.3) with $Y \rightsquigarrow X \times X$, $X \rightsquigarrow X \times X \times X$, $\mu \rightsquigarrow \gamma$ and $T \rightsquigarrow \pi^{13}$ yields

$$\begin{aligned}\int_{X \times X} d(x, y)^p \, d\sigma &= \int_{X \times X} d(x, y)^p \, d\left((\pi^{13})_{\#}\gamma\right)\\ &= \int_{X \times X \times X} d\left(\pi^{13}(x, y, z)\right)^p d\gamma(x, y, z).\end{aligned}$$

Since d satisfies the triangle inequality with a constant K, it follows from the definitions of projections that

$$\begin{aligned}d\left(\pi^{13}(x, y, z)\right) &= d(x, z) \le K\ (d(x, y) + d(y, z))\\ &= K\left(d\left(\pi^{12}(x, y, z)\right) + d\left(\pi^{23}(x, y, z)\right)\right).\end{aligned}$$

Therefore from the triangle inequality for the L^p-norm we obtain

$$\begin{aligned}&\left(\int_{X \times X \times X} d\left(\pi^{13}(x, y, z)\right)^p d\gamma(x, y, z)\right)^{1/p}\\ &\le K\left(\int_{X \times X \times X}\left(d\left(\pi^{12}(x, y, z)\right) + d\left(\pi^{23}(x, y, z)\right)\right)^p d\gamma(x, y, z)\right)^{1/p}\end{aligned}$$

$$
\begin{aligned}
&\leq K\left(\int_{X\times X\times X} d\left(\pi^{12}(x,y,z)\right)^p d\gamma(x,y,z)\right)^{1/p} \\
&\quad + K\left(\int_{X\times X\times X} d\left(\pi^{23}(x,y,z)\right)^p d\gamma(x,y,z)\right)^{1/p} \\
&= K\left(\int_{X\times X} d(x,y)^p\, d\left((\pi^{12})_{\#}\gamma\right)\right)^{1/p} \\
&\quad + K\left(\int_{X\times X} d(x,y)^p\, d\left((\pi^{23})_{\#}\gamma\right)\right)^{1/p} \qquad \text{again from (4.3)} \\
&= K\left(\int_{X\times X} d(x,y)^p\, d\gamma^{12}\right)^{1/p} + K\left(\int_{X\times X} d(x,y)^p\, d\gamma^{23}\right)^{1/p} \\
&= K\left(W_p(\mu_1,\mu_3) + W_p(\mu_3,\mu_2)\right).
\end{aligned}
$$

This completes the proof of the theorem. □

4.3 Topology Given by the Wasserstein Distance

We shall prove the following characterization.

Theorem 4.8 *Let X be a compact metric space. Then $W_1(\mu_n, \mu) \to 0$ as $n \to \infty$ if and only if $\mu_n \to \mu$ weakly, i.e., $\int_X f\, d\mu_n \to \int_X f\, d\mu$ for each $f \in C(X)$.*

Proof By Kantorovich-Rubinstein Theorem 6.9

$$
W_1(\mu_n,\mu) = \sup\left\{\int_X \phi\, d(\mu_n - \mu) : |\phi(x) - \phi(y)| \leq d(x,y) \quad \forall x, y \in X\right\}.
$$

We remark in passing that assuming this formula is trivial to show that W_1 satisfies the triangle inequality and disintegration of measures is not needed in such a case.

By Portmanteau's theorem, $\mu_n \to \mu$ weakly if and only if $\int_X f\, d\mu_n \to \int_X f\, d\mu$ for each f Lipschitz in X, i.e., with $|f(x) - f(y)| \leq K\, d(x,y)$ for some constant K, see, for example, [5, Remark 8.3.1] for a proof and comments (see also Exercise 20). Given such an f, we have that f/K and $-f/K$ are Lipschitz with constant one. So from Kantorovich-Rubinstein theorem

$$
\left|\int_X f\, d(\mu_n - \mu)\right| \leq K\, W_1(\mu_n,\mu).
$$

So if $W_1(\mu_n,\mu) \to 0$, then $\mu_n \to \mu$ weakly.

To prove the converse, let $L = \limsup_{n\to\infty} W_1(\mu_n, \mu)$ and we shall prove that $L = 0$. Indeed, there exists a subsequence μ_{n_k} such that

$$L = \lim_{k\to\infty} W_1(\mu_{n_k}, \mu).$$

From Kantorovich-Rubinstein, for each k there exists a function ϕ_{n_k} with Lipschitz constant bounded by one such that

$$W_1(\mu_{n_k}, \mu) \le \int_X \phi_{n_k}\, d(\mu_{n_k} - \mu) + \frac{1}{k}.$$

Since $\mu_{n_k}(X) = \mu(X) = 1$, it follows that $\int_X c\, d(\mu_{n_k} - \mu) = 0$ for any constant c. Then pick $x_0 \in X$ and define the functions $f_{n_k}(x) = \phi_{n_k}(x) - \phi_{n_k}(x_0)$. We have $|f_{n_k}(x)| \le d(x, x_0) \le C$ for all $x \in X$ and $|f_{n_k}(x) - f_{n_k}(y)| \le d(x, y)$. So the functions f_{n_k} are uniformly bounded and equicontinuous in X and by Arzela-Ascoli's theorem there is a subsequence $f_{n_{k_j}}$ converging uniformly in X to some function f. Clearly f is Lipschitz with constant bounded by one and

$$\int_X f_{n_{k_j}}\, d(\mu_{n_{k_j}} - \mu) = \int_X \phi_{n_{k_j}}\, d(\mu_{n_{k_j}} - \mu).$$

Hence

$$\begin{aligned}
L &\le \limsup_{j\to\infty} \left(\int_X f_{n_{k_j}}\, d(\mu_{n_{k_j}} - \mu) + \frac{1}{k_j} \right) \\
&\le \limsup_{j\to\infty} \int_X f_{n_{k_j}}\, d(\mu_{n_{k_j}} - \mu) \\
&= \limsup_{j\to\infty} \left(\int_X \left(f_{n_{k_j}} - f \right) d(\mu_{n_{k_j}} - \mu) + \int_X f\, d(\mu_{n_{k_j}} - \mu) \right) \\
&\le \limsup_{j\to\infty} \int_X \left(f_{n_{k_j}} - f \right) d(\mu_{n_{k_j}} - \mu) + \limsup_{j\to\infty} \int_X f\, d(\mu_{n_{k_j}} - \mu) \\
&\le \limsup_{j\to\infty} \| f_{n_{k_j}} - f \|_\infty \int_X d|\mu_{n_{k_j}} - \mu| = 0,
\end{aligned}$$

since μ_k, μ are probability measures and the uniform convergence. □

We also have the following.

Proposition 4.9 *Let X be a compact metric space and $1 \le p \le q < \infty$. Then $W_p \le W_q$ and $W_p \le C(X, p)\, W_1^{1/p}$. In particular, W_1 and W_p are comparable.*

Then from Theorem 4.8 and Proposition 4.9 we obtain the following.

Corollary 4.10 *Let X be a compact metric space and $1 \leq p < \infty$. Then $W_p(\mu_n, \mu) \to 0$ as $n \to \infty$ if and only if $\mu_n \to \mu$ weakly.*

Remark 4.11 The class of probability measures on a measure space is a convex set. Since $\mu(X) = \nu(Y) = 1$, the class $\Pi(\mu, \nu)$ is a convex subset of the probability measures over $X \times Y$. Then the Wasserstein measure satisfies

$$W_p((1-t)\mu_1 + t\mu_2, \nu) \leq (1-t)^{1/p}\, W_p(\mu_1, \nu) + t^{1/p}\, W_p(\mu_2, \nu)$$

for $0 \leq t \leq 1$ and for all μ_1, μ_2, ν probability Borel measures over X.

Remark 4.12 Let μ be a Borel probability measure in a compact space X and $x_0 \in X$. We shall prove that $\Pi(\mu, \delta_{x_0}) = \{\mu \otimes \delta_{x_0}\}$. It then follows that if $c : X \times X \to [0, \infty)$ is continuous we have

$$\min\left\{\int_{X\times X} c(x, y)\, d\gamma(x, y) : \gamma \in \Pi(\mu, \delta_{x_0})\right\} = \int_X c(x, x_0)\, d\mu(x).$$

In particular, if $\mu = \delta_{x_1}$ with $x_1 \in X$, it follows that

$$W_1\left(\delta_{x_1}, \delta_{x_0}\right) = d(x_1, x_0).$$

Indeed, let $\gamma \in \Pi(\mu, \delta_{x_0})$ we shall prove that $\gamma(A \times B) = \mu(A)\delta_{x_0}(B)$. If $x_0 \in B$, then

$$\gamma(A \times B) \leq \gamma(A \times X) = \mu(A) = \mu(A)\delta_{x_0}(B) = (\mu \otimes \delta_{x_0})(A \times B).$$

If $x_0 \notin B$, then

$$\gamma(A \times B) \leq \gamma(X \times B) = \delta_{x_0}(B) = 0 = (\mu \otimes \delta_{x_0})(A \times B),$$

so $\gamma(A \times B) = 0 = (\mu \otimes \delta_{x_0})(A \times B)$ if $x_0 \notin B$. Then in any case $\gamma(A \times B) \leq (\mu \otimes \delta_{x_0})(A \times B)$, and as a consequence

$$\gamma((X \setminus A) \times B) \leq (\mu \otimes \delta_{x_0})((X \setminus A) \times B).$$

It then remains to show that $\gamma(A \times B) \geq (\mu \otimes \delta_{x_0})(A \times B)$ when $x_0 \in B$. We have $1 = \delta_{x_0}(B) = \gamma(X \times B) \leq (\mu \otimes \delta_{x_0})(X \times B) = \mu(X) = 1$ when $x_0 \in B$. Hence

$$\begin{aligned} 1 &= \gamma(X \times B) = \gamma(((X \setminus A) \cup A) \times B) = \gamma(((X \setminus A) \times B) \cup (A \times B)) \\ &= \gamma((X \setminus A) \times B) + \gamma(A \times B) \end{aligned}$$

when $x_0 \in X$, and so

$$\begin{aligned}1 - \gamma(A \times B) &= \gamma((X \setminus A) \times B) \leq (\mu \otimes \delta_{x_0})((X \setminus A) \times B) \\ &= 1 - (\mu \otimes \delta_{x_0})(A \times B)\end{aligned}$$

obtaining

$$(\mu \otimes \delta_{x_0})(A \times B) \leq \gamma(A \times B)$$

when $x_0 \in X$.

Chapter 5
Multivalued Measure Preserving Maps

Abstract Multivalued measure preserving maps are introduced and a characterization of them is given.

Let (D, Σ, μ) and (D^*, Σ^*, μ^*) be measure spaces, and let $\mathcal{N} : D \to \mathcal{P}(D^*)$ be a multivalued map, i.e, $\mathcal{N}(x) \subseteq D^*$ for each $x \in D$. For $F \subseteq D^*$ denote

$$\mathcal{N}^{-1}(F) = \{x \in D : \mathcal{N}(x) \cap F \neq \emptyset\}.$$

We assume:

(1) $\mathcal{N}$ is measurable, i.e., $\mathcal{N}^{-1}(F) \in \Sigma$ for each $F \in \Sigma^*$;
(2) the set $\{x \in D : \mathcal{N}(x)$ is not a singleton$\}$ has μ-measure zero.

Definition 5.1 The push forward measure of μ through $\mathcal{N}$ is defined by

$$\mathcal{N}_{\#}\mu(F) = \mu\left(\mathcal{N}^{-1}(F)\right), \quad \text{for } F \in \Sigma^*.$$

Under the above assumptions on $\mathcal{N}$, we have that $\mathcal{N}_{\#}\mu$ is a measure in Σ^*. In fact, let $\{F_j\}_1^\infty$ be a sequence of disjoint sets in Σ^*. If $z \in \mathcal{N}^{-1}(F_i) \cap \mathcal{N}^{-1}(F_j)$ for $i \neq j$, then there exist $z_i \in \mathcal{N}(z) \cap F_i$ and $z_j \in \mathcal{N}(z) \cap F_j$. Since $F_i \cap F_j = \emptyset$, we have $z_i \neq z_j$. That is, $\mathcal{N}$ is not a singleton at z and by assumption (2) above we obtain that $\mu\left(\mathcal{N}^{-1}(F_i) \cap \mathcal{N}^{-1}(F_j)\right) = 0$ for $i \neq j$. This implies that $\mathcal{N}_{\#}\mu$ is σ-additive on Σ^*. In addition, $\mathcal{N}^{-1}(\emptyset) = \{x \in D : \mathcal{N}(x) \cap \emptyset \neq \emptyset\} = \emptyset$, and so $\mathcal{N}_{\#}\mu(\emptyset) = 0$. Therefore $\mathcal{N}_{\#}\mu$ is a measure in Σ^*.

Example 5.2 Suppose $u : \bar{\Omega} \to \mathbb{R}$ with u continuous, let $\mathcal{N} = \partial u$ be the sub differential of u in $\bar{\Omega}$ and let

$$\Sigma^* = \{E \subset \mathbb{R}^n : (\partial u)^{-1}(E) \text{ is a Lebesgue measurable subset of } \bar{\Omega}\}.$$

C. E. Gutiérrez, *Optimal Transport and Applications to Geometric Optics*,
SpringerBriefs on PDEs and Data Science,
https://doi.org/10.1007/978-981-99-4867-3_5

We shall prove that $\mathcal{N}$ satisfies (1) and (2) above with Σ the class of Lebesgue measurable subsets of $\bar{\Omega}$ and μ an absolutely continuous measure on $\bar{\Omega}$ with respect to Lebesgue measure; $\mu << |\cdot|$. Property (1) follows by definition of Σ^*, and property (2) follows from Lemma 2.7.

It remains to show that Σ^* is a σ-algebra containing the Borel sets of $\mathbb{R}^n$. We need to show that $F \in \Sigma^*$ implies $F^c \in \Sigma^*$; that Σ^* is closed by countable unions, and $(\partial u)^{-1}(K)$ is a closed set contained in $\bar{\Omega}$ for each $K \subset \mathbb{R}^n$ compact. Notice that if Ω is convex and u is convex in Ω, then $\mu\{x \in \bar{\Omega} : \mathcal{N}(x) = \emptyset\} = 0$.[1] Let $K \subset \mathbb{R}^n$ compact and let x_j be a sequence in $(\partial u)^{-1}(K)$ with $x_j \to x_0$. This means there exist $p_j \in \partial u(x_j) \cap K$. Since K is compact $p_j \to p_0 \in K$, by a subsequence, and so $x_j \to x_0$, by a subsequence. So $u(x) \geq u(x_0) + p_0 \cdot (x - x_0)$ for all $x \in \bar{\Omega}$, i.e., $p_0 \in \partial u(x_0) \cap K$, that is, $x_0 \in (\partial u)^{-1}(K)$. Since $\mathbb{R}^n = \cup_{j=1}^{\infty} K_j$ with K_j compact, and Σ^* is clearly closed by countable unions, we have that $\mathbb{R}^n \in \Sigma^*$. We also have the formula $(\partial u)^{-1}(\mathbb{R}^n \setminus F) = \left\{(\partial u)^{-1}(\mathbb{R}^n) \setminus (\partial u)^{-1}(F)\right\} \cup \left\{(\partial u)^{-1}(\mathbb{R}^n \setminus F) \cap (\partial u)^{-1}(F)\right\}$. By property (2), the second term in the last union has Lebesgue measure zero so we get that Σ^* is closed by complements.

Definition 5.3 We say the multivalued map $\mathcal{N}$ satisfying (1)–(2) above is measure preserving from μ to μ^* if $\mathcal{N}_{\#}\mu = \mu^*$.

Lemma 5.4 *Let (D, Σ, μ) and (D^*, Σ^*, μ^*) be finite measure spaces, and let $\mathcal{N} : D \to D^*$ be a multivalued map satisfying (1)–(2) above. Suppose in addition that D^* is a normal topological space, that is, any two disjoint closed sets can be separated by open neighborhoods, and μ^* is a regular Borel measure.*

If $\mathcal{N}$ is measure preserving from μ to μ^*, then

$$\int_D v(\mathcal{N}(x))d\mu(x) = \int_{D^*} v(y)\, d\mu^*(y), \tag{5.1}$$

for each $v \in C(D^*)$ bounded in D^*.

Conversely, if (5.1) *holds for each $v \in C(D^*)$ bounded, and*

$$\mu\{x \in D : \mathcal{N}(x) = \emptyset\} = 0, \tag{5.2}$$

then $\mathcal{N}$ is measure preserving from μ to μ^*.

[1] The boundary of a convex set in $\mathbb{R}^n$ has Lebesgue measure zero in $\mathbb{R}^n$. In fact, suppose 0 is in the interior of Ω and $p \in \partial\Omega$. There is an open ball $B_\epsilon(0) \subset \Omega$. By convexity the convex hull of p and $B_\epsilon(0)$ is contained in Ω, and the point λp is then in the interior of Ω for all $\lambda < 1$. We also have $\partial\Omega \subset (1/\lambda)\Omega$ for $\lambda < 1$. Now $\partial\Omega = \bar{\Omega} \setminus \Omega \subset (1/\lambda)\Omega \setminus \Omega$. So $|\partial\Omega| \leq |(1/\lambda)\Omega \setminus \Omega| = \left(\frac{1}{\lambda} - 1\right)|\Omega| \to 0$ as $\lambda \to 1^-$.

Proof Suppose $\mathcal{N}$ is measure preserving and let $v \in C(D^*)$. There is a sequence v_k of simple functions, i.e., $v_k(x) = \sum_{j=1}^{N_k} a_j^k \chi_{E_j^k}(x)$ with $\{E_j^k\}_{j=1}^{N_k}$ disjoint Borel sets in D^*, and $a_j^k \in \mathbb{R}$, such that, $v_k \to v$ uniformly in D^*. We have

$$\mu(\mathcal{N}^{-1}(E_j^k)) = \mu^*(E_j^k).$$

Now

$$\chi_{\mathcal{N}^{-1}(E)}(x) = \chi_E(\mathcal{N}(x))$$

for $E \subset D^*$ at each x such that $\mathcal{N}(x)$ is single valued, that is, a.e. in μ. Then

$$\begin{aligned}\mu(\mathcal{N}^{-1}(E)) &= \int_D \chi_{\mathcal{N}^{-1}(E)}(x)\, d\mu(x) = \int_D \chi_E(\mathcal{N}(x))\, d\mu(x) = \mu^*(E) \\ &= \int_{D^*} \chi_E(y)\, d\mu^*(y).\end{aligned}$$

Hence

$$\begin{aligned}\int_{D^*} v_k(y)\, d\mu^*(y) &= \int_{D^*} \sum_{j=1}^{N_k} a_j^k \chi_{E_j^k}(y)\, d\mu^*(y) = \sum_{j=1}^{N_k} a_j^k \int_{D^*} \chi_{E_j^k}(y)\, d\mu^*(y) \\ &= \sum_{j=1}^{N_k} a_j^k \int_D \chi_{E_j^k}(\mathcal{N}(x))\, d\mu(x) = \int_D v_k(\mathcal{N}(x))\, d\mu(x).\end{aligned}$$

Letting $k \to \infty$ we obtain (5.1).

To prove the converse, we first show that for any Borel set $E \subset D^*$

$$\mathcal{N}_\#\mu(E) \le \mu^*(E). \tag{5.3}$$

Indeed, let us assume first that $E = G$ is open, then given a compact set $K \subset G$, choose $v \in C(D^*)$ such that $0 \le v \le 1$, $v = 1$ on K, and $v = 0$ outside G. By (5.1), one gets

$$\begin{aligned}\mathcal{N}_\#\mu(K) = \mu(\mathcal{N}^{-1}(K)) &= \int_D \chi_{\mathcal{N}^{-1}(K)}(x)\, d\mu(x) = \int_D \chi_K(\mathcal{N}(x))\, d\mu(x) \\ &\le \int_D v(\mathcal{N}(x))\, d\mu(x) = \int_{D^*} v(y)\, d\mu^*(y) \le \mu^*(G),\end{aligned}$$

for each compact $K \subset G$. Since $\mathcal{N}_{\#}\mu$ is regular (because is a Borel measure finite on compacts), (5.3) follows for E open. For a general Borel set $E \subset D^*$, since μ^* is also regular, given $\epsilon > 0$ there exists G open $E \subset G$ with $\mu^*(G \setminus E) < \epsilon$. Then

$$\mathcal{N}_{\#}\mu(E) \leq \mathcal{N}_{\#}\mu(G) \leq \mu^*(G) = \mu^*(E) + \mu^*(G \setminus E) < \mu^*(E) + \epsilon$$

and so (5.3) follows.

We next prove that equality holds in (5.3).

First notice that

$$\{x \in D : \mathcal{N}(x) \neq \emptyset\} \cap (\mathcal{N}^{-1}(F))^c \subset \mathcal{N}^{-1}(F^c),$$

for any set $F \subset D^*$. Then applying (5.3) to $D^* \setminus F$ with F Borel set yields

$$\begin{aligned} \mu\left(\{x \in D : \mathcal{N}(x) \neq \emptyset\} \cap (\mathcal{N}^{-1}(F))^c\right) &\leq \mu\left(\mathcal{N}^{-1}(F^c)\right) \\ &= \mathcal{N}_{\#}\mu(F^c) \leq \mu^*(D^* \setminus F) \\ &= \mu^*(D^*) - \mu^*(F). \end{aligned}$$

Since from (5.2) $\mu\{x \in D : \mathcal{N}(x) = \emptyset\} = 0$, we have

$$\begin{aligned} \mu\left(\{x \in D : \mathcal{N}(x) \neq \emptyset\} \cap (\mathcal{N}^{-1}(F))^c\right) &= \mu\left((\mathcal{N}^{-1}(F))^c\right) \\ &= \mu(D) - \mu\left(\mathcal{N}^{-1}(F)\right) \\ &= \mu(D) - \mathcal{N}_{\#}\mu(F). \end{aligned}$$

So from the conservation condition $\mu(D) = \mu^*(D^*)$ (taking $v = 1$ in (5.1)), we obtain the reverse inequality in (5.3). □

Chapter 6
Kantorovich Dual Problem

Abstract The dual Kantorovich problem is introduced and it is proved, under apropppiate assumptions on the cost function, that this problem is equivalent to the Kantorovich and the Monge problems introduced in Chap. 1. It is also proved the invertibility of optimal maps for the Monge problem.

Consider a general Lipschitz cost function $c : D \times D^* \to \mathbb{R}$,[1] with D and D^* compact metric spaces, and *the set of admissible functions*

$$\mathcal{K} = \{(u, v) : u \in C(D), v \in C(D^*), u(x) + v(m) \leq c(x, m), \ \forall x \in D, \forall m \in D^*\}.$$

Suppose μ is a Borel measure in D and μ^* a Borel measure in D^*.[2] Define the *dual functional* I for $(u, v) \in C(D) \times C(D^*)$ by

$$I(u, v) = \int_D u(x)\, d\mu(x) + \int_{D^*} v(m)\, d\mu^*(m).$$

Notice that this is the dual of the Kantorovich problem in the following sense: if $\mu(D) = \mu^*(D^*) = 1$ and $\gamma \in \Pi(\mu, \mu^*)$, then the marginals of γ are μ and μ^* and so

$$\int_{D\times D^*} u(x)\, d\gamma(x, m) = \int_D u(x)\, d\mu(x), \quad \int_{D\times D^*} v(m)\, d\gamma(x, m) = \int_{D^*} v(m)\, d\mu^*(m),$$

[1] $|c(x_1, m_1) - c(x_2, m_2)| \leq K\,(d_D(x_1, x_2) + d_{D^*}(m_1, m_2))$.

[2] For the most part is needed that $\mu(D)$ and $\mu^*(D^*)$ are both finite.

C. E. Gutiérrez, *Optimal Transport and Applications to Geometric Optics*, SpringerBriefs on PDEs and Data Science,
https://doi.org/10.1007/978-981-99-4867-3_6

for each $u \in C(D)$ and $v \in C(D^*)$. Therefore

$$\begin{aligned}I(u, v) &= \int_D u(x)\, d\mu(x) + \int_{D^*} v(m)\, d\mu^*(m) \\ &= \int_{D\times D^*} u(x)\, d\gamma(x, m) + \int_{D\times D^*} v(m)\, d\gamma(x, m) \\ &= \int_{D\times D^*} (u(x) + v(m))\, d\gamma(x, m) \\ &\le \int_{D\times D^*} c(x, m)\, d\gamma(x, m).\end{aligned}$$

Hence $\sup_{(u,v)\in\mathcal{K}} I(u, v) \le \int_{D\times D^*} c(x, m)\, d\gamma(x, m)$ for each $\gamma \in \Pi(\mu, \mu^*)$ and so

$$\sup_{(u,v)\in\mathcal{K}} I(u, v) \le \inf\left\{\int_{D\times D^*} c(x, m)\, d\gamma(x, m) : \gamma \in \Pi(\mu, \mu^*)\right\}. \tag{6.1}$$

We then obtain that this is the continuous analogue of the primal and dual linear programming problems discussed in Sect. 1.9.

In this section we shall prove equality in (6.1).

Let us now define the *c and c^*- transforms*

$$\begin{aligned}u^c(m) &= \inf_{x\in D} [c(x, m) - u(x)], \quad m \in D^*; \\ v_c(x) &= \inf_{m\in D^*} [c(x, m) - v(m)], \quad x \in D.\end{aligned} \tag{6.2}$$

Definition 6.1 A function $\phi \in C(D)$ is c-concave if for $x_0 \in D$, there exist $m_0 \in D^*$ and $b \in \mathbb{R}$ such that $\phi(x) \le c(x, m_0) - b$ for each $x \in D$ with equality at $x = x_0$.

Definition 6.2 A function $\psi \in C(D^*)$ is c^*-concave if for $m_0 \in D^*$, there exist $x_0 \in D$ and $b \in \mathbb{R}$ such that $\psi(m) \le c(x_0, m) - b$ for each $m \in D^*$ with equality at $m = m_0$.

Obviously v_c is c-concave for any $v \in C(D^*)$, and v^c is c^*-concave for each $v \in C(D)$. We list the following useful properties:

(1) For any $u \in C(D)$ and $v \in C(D^*)$, $v_c \in Lip(D)$ and $u^c \in Lip(D^*)$ with Lipschitz constants bounded uniformly by the Lipschitz constant of c. Indeed, to show $u^c \in Lip(D^*)$, let $m_1, m_2 \in D^*$ and let $x_0 \in D$ be the point where the minimum in the definition of $u^c(m_2)$ is attained. We then have

$$\begin{aligned}u^c(m_1) - u^c(m_2) &= u^c(m_1) - (c(x_0, m_2) - u(x_0)) \\ &\le c(x_0, m_1) - u(x_0) - c(x_0, m_2) + u(x_0) \le K\, d_{D^*}(m_1, m_2).\end{aligned}$$

(2) If $(u, v) \in \mathcal{K}$, then $v(m) \le u^c(m)$ and $u(x) \le v_c(x)$. Also (v_c, v), $(u, u^c) \in \mathcal{K}$.
(3) $(\phi^c)_c \ge \phi$ for any $\phi : D \to \mathbb{R}$; and ϕ is c-concave iff $\phi = (\phi^c)_c$.

On one hand, from the definitions of c and c^* transforms we always have that $(\phi^c)_c \ge \phi$ for any ϕ.

On the other hand, if $\phi(x) \le c(x, m_0) - b$ in D and the equality holds at $x = x_0$, then $b = \phi^c(m_0)$. So $\phi(x_0) = c(x_0, m_0) - \phi^c(m_0)$ which yields $\phi(x_0) \ge (\phi^c)_c(x_0)$.
(4) $(\psi_c)^c \ge \psi$ for any $\psi : D^* \to \mathbb{R}$; and ψ is c^*-concave iff $\psi = (\psi_c)^c$.

As in the previous item, from the definitions of c and c^* transforms we always have that $(\psi_c)^c \ge \psi$ for any $\psi : D^* \to \mathbb{R}$.

On the other hand, if $\psi(m) \le c(x_0, m) - b$ in for all $m \in D^*$ and the equality holds at $m = m_0$, then $b = \psi_c(x_0)$. So $\psi(m_0) = c(x_0, m_0) - \psi_c(x_0)$ which yields $\psi(m_0) \ge (\psi_c)^c(m_0)$.

Definition 6.3 Given a function $\phi : D \to \mathbb{R}$, the c-normal mapping of ϕ is defined by

$$\mathcal{N}_{c,\phi}(x) = \{m \in D^* : \phi(x) + \phi^c(m) = c(x, m)\}, \qquad \text{for } x \in D,$$

and $\mathcal{T}_{c,\phi}(m) = \mathcal{N}_{c,\phi}^{-1}(m) = \{x \in D : m \in \mathcal{N}_{c,\phi}(x)\}$.

Notice that if $\phi : D \to \mathbb{R}$ is c-concave then $\mathcal{N}_{c,\phi}(x) \ne \emptyset$ for all $x \in D$.
Notice that from the definition of ϕ^c we have

$$\mathcal{N}_{c,\phi}(x) = \{m \in D^* : \phi(y) \le \phi(x) + c(y, m) - c(x, m) \,\forall\, y \in D\}.$$

We assume that the cost function $c(x, m)$ satisfies the following condition:

For any c-concave function ϕ,

the set $\{x \in D : \mathcal{N}_{c,\phi}(x)$ is not a singleton$\}$ has μ-measure zero, (6.3)

and $\mathcal{N}_{c,\phi}$ is measurable,

i.e., $(\mathcal{N}_{c,\phi})^{-1}(F)$ is μ-measurable for each $F \subseteq D^*$ 1Borel set.

Notice that if $c(x, m) = x \cdot m$, then $\mathcal{N}_{c,\phi}(x) = \partial^*\phi(x)$, where $\partial^*\phi$ is the super-differential of ϕ

$$\partial^*\phi(x) = \{m \in \mathbb{R}^n : \phi(y) \le \phi(x) + m \cdot (y - x) \,\forall\, y \in \Omega\},$$

and we have $\partial^*\phi(x) = -\partial(-\phi)(x)$. See [33] for large classes of costs satisfying (6.3). Consider the Monge cost $c(x, m) = |x - m|$ and let $D = D^* = B_1(0)$ be the unit ball in $\mathbb{R}^n$. If $\phi(x) = -|x|$, then ϕ is c-concave and the c-normal mapping is

$$\mathcal{N}_{c,\phi}(x) = \{\lambda x : 1 \le \lambda \le 1/|x|\}$$

for $x \neq 0$ and $\mathcal{N}_{c,\phi}(0) = B_1(0)$. Therefore, for Monge's cost $\mathcal{N}_{c,\phi}$ is multiple valued for all x in the interior of D.

Lemma 6.4 *Suppose that $c(x, m)$ satisfies the assumption* (6.3). *Then*

(i) *If ϕ is c-concave and $\mathcal{N}_{c,\phi}$ is measure preserving from μ to μ^*, then (ϕ, ϕ^c) is a maximizer of $I(u, v)$ in $\mathcal{K}$.*
(ii) *If ϕ is c-concave and (ϕ, ϕ^c) maximizes $I(u, v)$ in $\mathcal{K}$, then $\mathcal{N}_{c,\phi}$ is measure preserving from μ to μ^*.*

Proof We first prove (i). Given $(u, v) \in \mathcal{K}$, and since $\mathcal{N}_{c,\phi}$ is single valued except possibly on a set of μ-measure zero, we have

$$u(x) + v(\mathcal{N}_{c,\phi}(x)) \leq c(x, \mathcal{N}_{c,\phi}(x)) = \phi(x) + \phi^c(\mathcal{N}_{c,\phi}(x)), \qquad \text{a.e. } x \text{ on } D.$$

Integrating the above inequality with respect to μ, and since $\mathcal{N}$ is measure preserving we have from Lemma 5.4

$$\begin{aligned} I(u, v) &= \int_D u(x)\, d\mu(x) + \int_{D^*} v(m)\, d\mu^*(m) \\ &= \int_D u(x)\, d\mu(x) + \int_D v(\mathcal{N}_{c,\phi}(x))\, d\mu(x) \\ &\leq \int_D \phi(x)\, d\mu(x) + \int_D \phi^c(\mathcal{N}_{c,\phi}(x))\, d\mu(x) \\ &= \int_D \phi(x)\, d\mu(x) + \int_{D^*} \phi^c(m)\, d\mu^*(m) = I(\phi, \phi^c). \end{aligned}$$

From (2) above $(\phi, \phi^c) \in \mathcal{K}$ and the conclusion follows.

To prove (ii), let $\psi = \phi^c$, and for $v \in C(D^*)$, let $\psi_\theta(m) = \psi(m) + \theta\, v(m)$, where $\theta \in \mathbb{R}$; and let $\phi_\theta = (\psi_\theta)_c$. We shall prove that

$$\lim_{\theta \to 0} \frac{I(\phi_\theta, \psi_\theta) - I(\phi, \psi)}{\theta} = \int_D -v(\mathcal{N}_{c,\phi}(x))\, d\mu(x) + \int_{D^*} v(m)\, d\mu^*(m). \tag{6.4}$$

Since $(\phi_\theta, \psi_\theta) \in \mathcal{K}$, we have $I(\phi_\theta, \psi_\theta) \leq I(\phi, \psi)$ for $\theta \in \mathbb{R}$, and hence the existence of the limit (6.4) implies it must be zero. Therefore the measure preserving property of $\mathcal{N}_{c,\phi}$ follows from Lemma 5.4 since ϕ is c-concave and so (5.2) holds.

To prove (6.4) we write

$$\frac{I(\phi_\theta, \psi_\theta) - I(\phi, \psi)}{\theta} = \int_D \frac{\phi_\theta - \phi}{\theta}\, d\mu(x) + \int_{D^*} v(m)\, d\mu^*(m).$$

By Lebesgue dominated convergence theorem, to show (6.4), it is enough[3] to show that $\dfrac{\phi_\theta(x)-\phi(x)}{\theta}$ is uniformly bounded in D, for $|\theta|$ bounded, and $\dfrac{\phi_\theta(x)-\phi(x)}{\theta} \to -v(\mathcal{N}_{c,\phi}(x))$ for all $x \in D \setminus S$, where $\mathcal{N}_{c,\phi}(x)$ is single-valued on $D \setminus S$ and $\mu(S) = 0$. Let us first prove the uniform boundedness. Fix $x \in D$, we have by continuity that $\phi_\theta(x) = c(x, m_\theta) - \psi_\theta(m_\theta)$ for some $m_\theta \in D^*$. Since ϕ is c-concave there exists $m_1 \in D^*$ and $b \in \mathbb{R}$ such that $\phi(y) \le c(y, m_1) - b$ for all $y \in D$ with equality when $y = x$. This implies that $b = \phi^c(m_1)$ and so $\phi(x) = c(x, m_1) - \psi(m_1)$. Hence

$$\begin{aligned}\phi_\theta(x) - \phi(x) &= c(x, m_\theta) - \psi(m_\theta) - \theta\, v(m_\theta) - \phi(x)\\ &\ge \psi_c(x) - \theta\, v(m_\theta) - \phi(x) = (\phi^c)_c(x)\\ &\quad - \theta\, v(m_\theta) - \phi(x) \ge -\theta\, v(m_\theta),\end{aligned}$$

by (3) above. We also have

$$\begin{aligned}\phi_\theta(x) - \phi(x) &= \phi_\theta(x) - c(x, m_1) + \psi(m_1)\\ &= \phi_\theta(x) - c(x, m_1) + \psi_\theta(m_1) - \theta\, v(m_1)\\ &\le \phi_\theta(x) - (\psi_\theta)_c(x) - \theta\, v(m_1) = -\theta\, v(m_1).\end{aligned}$$

Then we get

$$-\theta\, v(m_\theta) \le \phi_\theta(x) - \phi(x) \le -\theta\, v(m_1),$$

with $m_1, m_\theta \in D^*$ depending on x. Moreover, if $x \in D \setminus S$, then $m_1 = \mathcal{N}_{c,\phi}(x)$ since $\psi = \phi^c$. To finish the proof, we show that m_θ converges to m_1 as $\theta \to 0$. Otherwise, there exists a sequence m_{θ_k} such that $m_{\theta_k} \to m_\infty \ne m_1$. So $\phi(x) = \lim_{\theta\to 0} \phi_\theta(x) = c(x, m_\infty) - \psi(m_\infty)$, which yields $m_\infty \in \mathcal{N}_{c,\phi}(x)$. We then get $m_1 = m_\infty$, a contradiction. The proof is complete. □

Lemma 6.5 *Suppose $\mu(D) = \mu^*(D^*)$. There exists ϕ c-concave such that*

$$I(\phi, \phi^c) = \sup\{I(u, v) : (u, v) \in \mathcal{K}\}.$$

Proof Let

$$I_0 = \sup\{I(u, v) : (u, v) \in \mathcal{K}\},$$

and let $(u_k, v_k) \in \mathcal{K}$ be a sequence such that $I(u_k, v_k) \to I_0$. Set $\bar{u}_k = (v_k)_c$ and $\bar{v}_k = (\bar{u}_k)^c$. From property (2) above, since $(u_k, v_k) \in \mathcal{K}$, we have $(\bar{u}_k, v_k) =$

[3] We are using here that $\mu(D) < \infty$.

$((v_k)_c, v_k) \in \mathcal{K}$, which again by property (2) implies $(\bar{u}_k, \bar{v}_k) \in \mathcal{K}$. We also have by property (2), $u_k \le \bar{u}_k$, and $v_k \le \bar{v}_k$, and therefore $I(\bar{u}_k, \bar{v}_k) \to I_0$. Let $c_k = \min_D \bar{u}_k = \bar{u}_k(x_k)$ (for some $x_k \in D$) and define

$$u_k^\sharp(x) = \bar{u}_k(x) - c_k, \qquad v_k^\sharp(m) = \bar{v}_k(m) + c_k.$$

By property (1), $\bar{u}_k$ and $\bar{v}_k$ are Lipschitz with a constant bounded by the Lipschitz constant of c, and so the same holds for $u_k^\sharp$ and $v_k^\sharp$. Obviously, $(u_k^\sharp, v_k^\sharp) \in \mathcal{K}$ and since $\mu(D) = \mu^*(D^*)$, we have $I(\bar{u}_k, \bar{v}_k) = I(u_k^\sharp, v_k^\sharp)$. By property (1) above, $\bar{u}_k$ are uniformly Lipschitz, and so $u_k^\sharp$ are uniformly bounded for all k. In addition, $v_k^\sharp = (\bar{u}_k)^c + c_k = (u_k^\sharp)^c$ and consequently $v_k^\sharp$ are also uniformly bounded for all k. By Arzelá-Ascoli's theorem, $(u_k^\sharp, v_k^\sharp)$ contains a subsequence converging uniformly to (ϕ, ψ) in $D \times D^*$. We then obtain that $(\phi, \psi) \in \mathcal{K}$ and $I_0 = \sup\{I(u, v) : (u, v) \in \mathcal{K}\} = I(\phi, \psi)$. Notice that this shows in particular that the supremum of I over $\mathcal{K}$ is finite. From property (2) above, $(\psi_c, \psi) \in \mathcal{K}$ and $\phi \le \psi_c$, so $I_0 = I(\phi, \psi) \le I(\psi_c, \psi)$ and we get $I_0 = I(\psi_c, \psi)$. Once again from property (2) above, $(\psi_c, (\psi_c)^c) \in \mathcal{K}$ and $\psi \le (\psi_c)^c$, and so $I_0 = I(\psi_c, \psi) \le I(\psi_c, (\psi_c)^c)$ and we get $I_0 = I(\psi_c, (\psi_c)^c)$. Therefore $(\psi_c, (\psi_c)^c)$ is the sought maximizer of $I(u, v)$, and ψ_c is c-concave; and from Item 1 above is Lipschitz. □

6.1 Kantorovich dual = Monge = Kantorovich

We now prove these problems all give the same value.

Lemma 6.6 *Suppose that $c(x, m)$ satisfies the assumption (6.3). Let $(\phi, \phi^c) \in \mathcal{K}$, with ϕ c-concave, be a maximizer of $I(u, v)$ in $\mathcal{K}$. Then $\inf_{s \in \mathcal{S}} \int_D c(x, s(x))\, d\mu(x)$ is attained at $s = N_{c,\phi}$, where $\mathcal{S}$ is the class of measure preserving mappings from μ to μ^*. Moreover*

$$\inf_{s \in \mathcal{S}} \int_D c(x, s(x))\, d\mu(x) = \sup\{I(u, v) : (u, v) \in \mathcal{K}\}. \tag{6.5}$$

Proof Let $\psi = \phi^c$. For $s \in \mathcal{S}$, we have, since $(\phi, \phi^c) \in \mathcal{K}$, that

$$\begin{aligned}
\int_D c(x, s(x))\, d\mu(x) &\ge \int_D (\phi(x) + \psi(s(x)))\, d\mu(x) \\
&= \int_D \phi(x)\, d\mu(x) + \int_D \psi(s(x))\, d\mu(x) \\
&= \int_D \phi(x)\, d\mu(x) + \int_{D^*} \psi(m)\, d\mu^*(m) = I(\phi, \psi)
\end{aligned}$$

$$= \int_D \big(\phi(x) + \psi(\mathcal{N}_{c,\phi}(x))\big)\, d\mu(x), \text{ from Lemma 6.4(ii)}$$

$$= \int_D c(x, \mathcal{N}_{c,\phi}(x))\, d\mu(x).$$

□

Lemma 6.7 *Suppose that $c(x, m)$ satisfies the assumption (6.3), and $\mu(G) > 0$ for any non empty open set $G \subset D$; i.e., μ is strictly positive. Then the mapping minimizing $\inf_{s \in \mathcal{S}} \int_D c(x, s(x))\, d\mu(x)$ is unique in the class of measure preserving mappings from μ to μ^* satisfying (1)–(2) from Sect. 5.*

Proof From Lemmas 6.5 and 6.6, let $\mathcal{N}_{c,\phi}$ be a minimizing mapping associated with a maximizer (ϕ, ϕ^c) of $I(u, v)$ with ϕ c-concave. Suppose that $\mathcal{N}_0$ is another minimizing mapping that is measure preserving. From Lemmas 5.4 and 6.6 we have

$$\int_D \big(c(x, \mathcal{N}_0(x)) - \phi(x) - \phi^c(\mathcal{N}_0(x))\big)\, d\mu(x)$$

$$= \inf_{s \in \mathcal{S}} \int_D c(x, s(x))\, d\mu(x) - \left(\int_D \phi(x)\, d\mu(x) + \int_{D^*} \phi^c(m)\, d\mu^*(m)\right) = 0.$$

Since $\phi(x) + \phi^c(\mathcal{N}_0(x)) \le c(x, \mathcal{N}_0(x))$, it follows, from the assumption on the measure μ, that $\phi(x) + \phi^c(\mathcal{N}_0(x)) = c(x, \mathcal{N}_0(x))$ for all $x \in D \setminus S$ with $\mu(S) = 0$. By definition of $\mathcal{N}_{c,\phi}$, this means $\mathcal{N}_0(x) \in \mathcal{N}_{c,\phi}(x)$ for each $x \in D \setminus S$. From (6.3), $\mathcal{N}_{c,\phi}(x)$ is a singleton for each $x \in D \setminus S'$ with $\mu(S') = 0$. Therefore for each $x \in D \setminus (S \cup S')$ we have $\mathcal{N}_0(x) = \mathcal{N}_{c,\phi}(x)$. That is, $\mathcal{N}_0$ and $\mathcal{N}_{c,\phi}$ are equal except possibly on a set of μ-measure zero. □

We then conclude from (1.3), (6.1), and Lemma 6.6 that the Monge problem, the Kantorovich problem, and the dual Kantorovich problem are "all equal," meaning that the chain of identities in the following theorem hold.

Theorem 6.8 (Strong Duality) *We have*

$$\min\left\{\int_{X \times Y} c(x, y)\, d\gamma : \gamma \in \Pi(\mu, \nu)\right\}$$

$$= \inf\left\{\int_X c\,(x, Tx)\; d\mu : T \in \mathcal{S}(\mu, \nu)\right\}$$

$$= \max\left\{\int_X u(x)\, d\mu(x) \right.$$

$$\left. + \int_Y v(m)\, d\nu(m) : (u, v) \in C(X) \times C(Y), u(x) + v(m) \le c(x, m)\right\}$$

when the cost c is Lipschitz and satisfies condition (6.3)*, and X, Y are compact metric spaces with $\mu(X) = \nu(Y) = 1$.*

We conclude from this theorem the following result used in the proof of Theorem 4.8.

Theorem 6.9 (Kantorovich-Rubinstein) *If (D, d) is a compact metric space, μ, ν are Borel probability measures in D, and c is Lipschitz and satisfies condition* (6.3)*, then*

$$\sup\left\{\int_D f\,d\mu - \int_D f\,d\nu : |f(x) - f(y)| \leq d(x, y) \quad \forall x, y \in D\right\}$$
$$= \min\left\{\int_{D\times D} d(x, y)\,d\gamma(x, y) : \gamma \in \Pi(\mu, \nu)\right\}.$$

Proof Let $(u, v) \in \mathcal{K} = \{(u, v) : u, v \in C(D), u(x) + v(y) \leq c(x, y), \forall x, y \in D\}$ and set $g(x) = \inf_{y\in D}(d(x, y) - v(y))$. Then $|g(x) - g(y)| \leq d(x, y)$ for all $x, y \in D$, $u(x) \leq g(x)$, and $v(x) \leq -g(x)$. Hence

$$\int_D u\,d\mu + \int_D v\,d\nu \leq \sup\left\{\int_D f\,d\mu - \int_D f\,d\nu : |f(x) - f(y)| \right.$$
$$\left. \leq d(x, y) \quad \forall x, y \in D\right\}.$$

On the other hand, if $\gamma \in \Pi(\mu, \nu)$, then $\int_D f(x)\,d\mu(x) = \int_{D\times D} f(x)\,d\gamma(x, y)$, $\int_D f(x)\,d\nu(x) = \int_{D\times D} f(y)\,d\gamma(x, y)$ for each $f \in C(D)$, and we obtain

$$\sup\left\{\int_D f\,d\mu - \int_D f\,d\nu : |f(x) - f(y)| \leq d(x, y) \quad \forall x, y \in D\right\}$$
$$\leq \int_{D\times D} d(x, y)\,d\gamma(x, y)$$

for each $\gamma \in \Pi(\mu, \nu)$. Taking supremum over $(u, v) \in \mathcal{K}$ and applying the strong duality Theorem 6.8 the desired result follows. □

6.1.1 Invertibility of Optimal Maps

If $\mathbf{r}$ is the optimal map for Monge's problem for the cost c transporting (D, μ) into (D^*, μ^*), we analyze here whether the optimal map $\mathbf{s}$ for the cost $\tilde{c}$ transporting (D^*, μ^*) into (D, μ) is the inverse of $\mathbf{r}$, and viceversa. Recall that c is a Lipschitz cost $c : D \times D^* \to \mathbb{R}_{\geq 0}$, and $\tilde{c} : D^* \times D \to \mathbb{R}_{\geq 0}$ is given by $\tilde{c}(m, x) = c(x, m)$.

The function $\psi : D^* \to \mathbb{R}$ is $\tilde{c}$-concave if for each $m_0 \in D^*$ there exist $x_0 \in D$ and $b \in \mathbb{R}$ such that $\psi(m) \leq \tilde{c}(m, x_0) - b$ for all $m \in D^*$ with equality at

$m = m_0$. It is clear that ψ is $\tilde{c}$-concave if and only if ψ is c^*-concave in the sense of Definition 6.2 and therefore $\psi = (\psi_c)^c$ from Item 4 above. Also from the definitions in (6.2), $\psi^{\tilde{c}}(x) = \psi_c(x)$ for $x \in D$.

Given $\psi : D^* \to \mathbb{R}$, consider now the $\tilde{c}$-subdifferential given by

$$\mathcal{N}_{\tilde{c},\psi}(m) = \left\{x \in D : \psi(m) + \psi^{\tilde{c}}(x) = \tilde{c}(m,x)\right\}, \quad m \in D^*.$$

For each c^*-concave function ψ, we have that $\psi(m) + \psi^{\tilde{c}}(x) = \tilde{c}(m,x)$ if and only if $\psi^c(x) + \psi(m) = c(x,m)$ implying that

$$\left(\mathcal{N}_{c,\psi_c}\right)^{-1}(m) = \mathcal{N}_{\tilde{c},\psi}(m), \quad \text{for each } m \in D^*. \tag{6.6}$$

In order to apply the existence and uniqueness results from Sect. 6.1 to obtain the optimal map for the cost $\tilde{c}$ and transporting (D^*, μ^*) to (D, μ), we assume the following condition (similar to (6.3)):

For any c^*-concave function ψ,

$$\textit{the set } \{m \in D^* : \mathcal{N}_{\tilde{c},\psi}(m) \text{ is not a singleton}\} \textit{ has } \mu^*\textit{-measure zero}, \tag{6.7}$$

and $\mathcal{N}_{\tilde{c},\psi}$ is measurable,

i.e., $(\mathcal{N}_{\tilde{c},\psi})^{-1}(F)$ is μ^*-measurable for each $F \subseteq D$ Borel set.

The analogue of the class $\mathcal{K}$ for $\tilde{c}$ is clearly

$$\tilde{\mathcal{K}} = \{(u', v') \in C(D^*) \times C(D) : u'(m) + v'(x) \leq \tilde{c}(m,x)\},$$

and $(u, v) \in \mathcal{K}$ if and only if $(v, u) \in \tilde{\mathcal{K}}$. Also, if we set $J(u', v') = \int_{D^*} u'(m)\, d\mu^*(m) + \int_D v'(x)\, d\mu(x)$ for $(u', v') \in \tilde{\mathcal{K}}$, then $I(u, v) = J(v, u)$ for $(u, v) \in \mathcal{K}$. Consequently,

$$\max_{(u,v)\in\mathcal{K}} I(u,v) = \max_{(u',v')\in\tilde{\mathcal{K}}} J(u',v').$$

Since c satisfies (6.3), it then follows from Lemma 6.5 that there exists ϕ c-concave such that (ϕ, ϕ^c) maximizes I over $\mathcal{K}$. Then from Lemmas 6.6 and 6.7, $\mathcal{N}_{c,\phi}$ is the unique optimal map for the Monge problem with the cost c transporting (D, μ) into (D^*, μ^*) (of course, for the uniqueness we assume $\mu(G) > 0$ for each open set).

Assuming (6.7) and that $\mu^*(G) > 0$ for each open set, let us now apply Lemmas 6.6 and 6.7 to the cost $\tilde{c}$. From Lemma 6.5, there exists $\psi : D^* \to \mathbb{R}$ c^*-concave such that $\left(\psi, \psi^{\tilde{c}}\right) (= (\psi, \psi_c))$ maximizes J over $\tilde{\mathcal{K}}$, and so $\mathcal{N}_{\tilde{c},\psi}$ is the unique optimal map for the Monge problem with cost $\tilde{c}$ transporting (D^*, μ^*) into (D, μ). Moreover, if (ψ, ψ_c) maximizes J over $\tilde{\mathcal{K}}$, then (ψ_c, ψ) maximizes I over $\mathcal{K}$. Therefore, since ψ_c is c-concave, once again by Lemmas 6.6 and 6.7, we get that

$\mathcal{N}_{c,\psi_c}$ is the unique optimal map for the Monge problem with the cost c transporting (D, μ) into (D^*, μ^*). Hence

$$\mathcal{N}_{c,\psi_c}(x) = \mathcal{N}_{c,\phi}(x), \quad \mu\text{-a.e. in } x \in D, \tag{6.8}$$

and from (6.6) we obtain

$$\mathcal{N}_{\tilde{c},\psi} = \left(\mathcal{N}_{c,\phi}\right)^{-1}. \tag{6.9}$$

That is, the optimal map backwards is the inverse of the optimal map forward.

Let us see the pointwise invertibility. Set $\mathbf{r}(x) = \mathcal{N}_{c,\phi}(x)$, $x \in D$ and $\mathbf{s}(m) = \mathcal{N}_{\tilde{c},\psi}(m)$, $m \in D^*$. From (6.3), there exists $E_1 \subset D$ with $\mu(D \setminus E_1) = 0$ such that $\mathbf{r}(x)$ is a single point for each $x \in E_1$. Since ψ_c is c-concave, again from (6.3) there exists $E_2 \subset D$ with $\mu(D \setminus E_2) = 0$ such that $\mathcal{N}_{c,\psi_c}(x)$ is a single point for each $x \in E_2$. From (6.8), there is $E_3 \subset D$ with $\mu(D \setminus E_3) = 0$ such that $\mathcal{N}_{c,\psi_c}(x) = \mathbf{r}(x)$ for each $x \in E_3$. If $E = E_1 \cap E_2 \cap E_3$, then $\mu(D \setminus E) = 0$. Writing $D^* = \mathbf{s}^{-1}(D) = \mathbf{s}^{-1}(D \setminus E) \cup \mathbf{s}^{-1}(E)$, since $\mathbf{s}$ is measure preserving (D^*, μ^*) to (D, μ), it follows that $\mu^*\left(\mathbf{s}^{-1}(D \setminus E)\right) = \mu(D \setminus E) = 0$. So if we let $H = \mathbf{s}^{-1}(E)$, then $\mu^*(D^* \setminus H) = 0$. Let $m \in H$. Then $\mathbf{s}(m) \in E$, so $\mathcal{N}_{c,\psi_c}(\mathbf{s}(m)) = \mathbf{r}(\mathbf{s}(m))$ since $\mathbf{s}(m) \in E_3$. Since $\mathbf{s}(m) \in E_2$, $\mathbf{r}(\mathbf{s}(m))$ is the only point in $\mathcal{N}_{c,\psi_c}(\mathbf{s}(m))$. But from (6.6) $m \in \mathcal{N}_{c,\psi_c}(\mathbf{s}(m))$. Therefore, $m = \mathbf{r}(\mathbf{s}(m))$ for each $m \in H$, i.e., μ^*-a.e. in D^*. Moreover, since the whole argument above is symmetric, we obtain also that $\mathbf{s}(\mathbf{r}(x)) = x$, μ-a.e. in D.

Chapter 7
Brenier and Aleksandrov Solutions

Abstract The notions of Alexandrov and Brenier solutions for the subdifferential of a convex function are introduced. It is proved that each Alexandrov solution is a Brenier solution but the converse does not hold in general.

Let $f \in L^1(D)$ and $g \in L^1(D^*)$, where D, D^* are domains in $\mathbb{R}^n$ satisfying $\int_D f(x)\,dx = \int_{D^*} g(y)\,dy$. If u is convex in D and solves the problem

$$\int_{(\partial u)^{-1}(E)} f(x)\,dx = \int_E g(y)\,dy \qquad \forall E \subset D^* \text{ Borel set} \tag{7.1}$$

that is, the sub differential ∂u is measure preserving from $f\,dx$ to $g\,dy$ in the sense of Definition 5.3, then we say that u is a Brenier solution.

If on the other hand, v is convex in D and solves

$$\int_F f(x)\,dx = \int_{\partial v(F)} g(y)\,dy \qquad \forall F \subset D \text{ Borel set}; \tag{7.2}$$

then we say that v is an Aleksandrov solution.

We shall prove that if u satisfies (7.2), then u satisfies (7.1); will show later that the converse is not true. To do this we shall use that $y \in \partial u(x)$ if and only if $x \in \partial u^*(y)$ where u^* is the Legendre transform of u, and define the following sets:

$$G = \{x \in D : u \text{ is differentiable at } x\}$$

$$M = \{y \in \partial u(G) : u^* \text{ is differentiable at } y\}$$

$$F = (\partial u)^{-1}(M) \cap G.$$

We shall prove first that

(a) $\partial u(F) = M$;
(b) $\partial u : F \to M$ is bijective;

C. E. Gutiérrez, *Optimal Transport and Applications to Geometric Optics*, SpringerBriefs on PDEs and Data Science,
https://doi.org/10.1007/978-981-99-4867-3_7

(c) for each $E \subset D^*$ we have $|(\partial u)^{-1}(E \setminus M)| = 0$.

Let us prove (a): if $y \in \partial u(F)$, then there is $x \in F$ such that $y \in \partial u(x)$. Since $x \in F$, then $x \in G$ and so u is differentiable at x, so $y = \nabla u(x) = \partial u(x)$. Also since $x \in (\partial u)^{-1}(M)$, $\partial u(x) \cap M \neq \emptyset$ and so $y \in M$. On the other hand, if $y \in M$, then $y \in \partial u(x)$ for some $x \in G$ and u^* is differentiable at y. So u is differentiable at x and $y = \nabla u(x) = \partial u(x)$, that is, $y \in \partial u(x) \cap M$ so by definition $x \in (\partial u)^{-1}(M)$. So $x \in F$.

Proof of (b): the surjectivity is part (a). Let $x_1, x_2 \in F$ with $p = \nabla u(x_1) = \nabla u(x_2) \in M$. Then $x_1, x_2 \in \partial u^*(p)$, and since u^* is differentiable at p, we get $x_1 = x_2$.

Proof of (c): Set $S = (\partial u)^{-1}(E \setminus M)$. We can write $S = (S \cap G) \cup (S \cap G^c)$, and since $|G^c| = 0$ is enough to show that $|S \cap G| = 0$. If $x \in S \cap G$, then u is differentiable at x and $\partial u(x) \cap (E \setminus M) \neq \emptyset$. Take any $p \in \partial u(x) \cap (E \setminus M)$. That is, $p \in \partial u(x) = \nabla u(x)$, $p \in E$ and $p \notin M$. Since $x \in G$, $p \in \partial u(G)$. Also $x \in \partial u^*(p)$. Now

$$M = (\partial u(G)) \cap \{y : u^* \text{ is differentiable at } y\},$$

so

$$M^c = (\partial u(G))^c \cup \{y : u^* \text{ is not differentiable at } y\}.$$

Hence u^* is not differentiable at p and so there is $x_0 \neq x$ such that $x_0 \in \partial u^*(p)$ implying $p \in \partial u(x_0)$. Therefore $p \in \partial u(x_0) \cap \partial u(x)$. We have then proved that

$$\nabla u(x) \in \{p : \exists x_1 \neq x_2, p \in \partial u(x_1) \cap \partial u(x_2)\} = S_0$$

for each $x \in S \cap G$. From Lemma 2.7 $|S_0| = 0$, we then get that $\nabla u(S \cap G) = \partial u(S \cap G)$ has measure zero. If u satisfies (7.2), then

$$\int_{S\cap G} f(x)\, dx = \int_{\partial u(S\cap G)} g(y)\, dy = 0$$

and if f is positive a.e., we then obtain that $|S \cap G| = 0$, and (c) is proved.

Let us then prove that u is a Brenier solution. We write

$$\begin{aligned}
&\int_{(\partial u)^{-1}(E)} f(x)\, dx \\
&= \int_{(\partial u)^{-1}(E\cap M)\cup(\partial u)^{-1}(E\setminus M)} f(x)\, dx \\
&= \int_{(\partial u)^{-1}(E\cap M)} f(x)\, dx
\end{aligned}$$

by (c) and since $|(\partial u)^{-1}(E \cap M) \cap (\partial u)^{-1}(E \setminus M)| = 0$

$$= \int_{\partial u((\partial u)^{-1}(E\cap M))} g(y)\,dy \qquad \text{by (7.2)}$$

$$= \int_{E\cap M} g(y)\,dy \qquad \text{by (b)}$$

$$= \int_E g(y)\,dy \qquad \text{since } |E \setminus M| = 0.$$

Let us now construct a Brenier solution that is not an Aleksandrov solution. Consider in dimension two $D = B_1(0)$, $D^* = B_{\sqrt{2}}(0) \setminus B_1(0)$, $f = g = 1$, and $\varphi(t) = \int_0^t \sqrt{1+r^2}\,dr$, with $t \geq 0$. Let

$$u(x) = \varphi(|x|).$$

Since φ is increasing and convex in $[0, +\infty)$, the function u is convex. For a general function $\psi \in C^2(0, +\infty)$, and $v(x) = \psi(|x|)$ with $x \in \mathbb{R}^n$, from the Sherman-Morrison determinant formula it follows that

$$\det D^2 v = \left(\frac{\psi'(|x|)}{|x|}\right)^{n-1} \psi''(|x|).$$

Therefore, with $n = 2$ taking $\psi = \varphi$ we have $\det D^2 u = 1$ in $B_1(0) \setminus 0$. We also have $Du(x) = \dfrac{\sqrt{1+|x|^2}}{|x|}x$ for $x \neq 0$, so $Du(B_1(0) \setminus 0) = D^*$. For $E \subset D^*$, it follows that $\int_E dy = \int_{(Du)^{-1}(E)} \det D^2u(x)\,dx = |(Du)^{-1}(E)|$, and so u satisfies (7.1). On the other hand, $u(x) \geq |x| \geq p \cdot x$ for all $|x| \leq 1$ and $|p| \leq 1$, which implies $B_1(0) \subset \partial u(0)$. If $p \in \partial u(0)$, then $u(x) \geq p \cdot x$ for all $|x| \leq 1$ and taking $x = \epsilon\, p/|p|$ and letting $\epsilon \to 0$ yields $|p| \leq 1$. Therefore $B_1(0) = \partial u(0)$. We have for $F \subset B_1(0)$ that $\partial u(F) = \partial u(F \setminus 0) \cup \partial u(F \cap 0)$, therefore

$$|\partial u(F)| = |\partial u(F \setminus 0)| + |\partial u(F \cap 0)| = \int_{\partial u(F\setminus 0)} dy + |\partial u(F \cap 0)|$$

$$= \int_{F\setminus 0} \det D^2u(x)\,dx + |\partial u(F \cap 0)| = |F| + |\partial u(F \cap 0)|$$

which means that u is a Aleksandrov solution to the Monge-Ampère equation $Mu = 1 + \pi\,\delta_0$; that is, u does not solve (7.2).

Remark 7.1 If the domain D^* is convex, then each Brenier solution is an Aleksandrov solution, see [9].

Chapter 8
Cyclical Monotonicity

Abstract In this chapter the notion of cyclical monotonicity of multi-valued maps is analyzed.

Definition 8.1 A multivalued map $s : \mathbb{R}^n \to \mathbb{R}^n$ is cyclically monotone if for each non-negative integer N and for any points $x_0, \cdots, x_N$ and $y_0, \cdots, y_N$ with $y_i \in s(x_i)$, $0 \le i \le N$, we have

$$\begin{aligned}&(x_1 - x_0) \cdot y_0 + (x_2 - x_1) \cdot y_1 + \cdots + (x_{N-1} - x_{N-2}) \cdot y_{N-2}\\&+ (x_N - x_{N-1}) \cdot y_{N-1} + (x_0 - x_N) \cdot y_N \le 0.\end{aligned}$$

If s_1 and s_2 are two multivalued maps then we say that $s_1 \subseteq s_2$ if $s_1(x) \subseteq s_2(x)$ for all x. The graph of the multivalued map s is given by

$$G(s) = \{(x, y) \in \mathbb{R}^n \times \mathbb{R}^n : y \in s(x)\}.$$

A multivalued map $s : \mathbb{R}^n \to \mathbb{R}^n$ is *maximal cyclically monotone* if $G(s)$ is not contained in the graph of any other cyclically monotone map from $\mathbb{R}^n$ to $\mathbb{R}^n$.

Theorem 8.2 ([48]) *If $s : \mathbb{R}^n \to \mathbb{R}^n$ is cyclically monotone, then there exists a convex function $f : \mathbb{R}^n \to (-\infty, +\infty]$, not identically $+\infty$, such that $s \subset \partial f$. Reciprocally, if f is convex, then ∂f is cyclically monotone. Therefore, if s is maximal cyclically monotone, then $s = \partial f$. If s is single valued a.e., then $s = \partial f$ a.e.*

Proof Fix any pair (x_0, y_0) with $y_0 \in s(x_0)$, and for each $x \in \mathbb{R}^n$ define the function

$$f(x) = \sup\{(x - x_N) \cdot y_N + (x_N - x_{N-1}) \cdot y_{N-1} + \cdots + (x_2 - x_1) \cdot y_1 + (x_1 - x_0) \cdot y_0\}$$

where the supremum is taken over all $N \ge 0$ and all pairs (x_i, y_i) with $y_i \in s(x_i)$, $0 \le i \le N$. Since the supremum is taken over affine functions, it follows that the

C. E. Gutiérrez, *Optimal Transport and Applications to Geometric Optics*,
SpringerBriefs on PDEs and Data Science,
https://doi.org/10.1007/978-981-99-4867-3_8

function f is convex. Since s is cyclically monotone, if $N \geq 1$ we have at $x = x_0$ that

$$(x_0 - x_N) \cdot y_N + (x_N - x_{N-1}) \cdot y_{N-1} + \cdots + (x_2 - x_1) \cdot y_1 + (x_1 - x_0) \cdot y_0 \leq 0.$$

If $N = 0$, then $(x - x_0) \cdot y_0$ equals zero when $x = x_0$. Therefore, $f(x_0) = 0$, and so f is not identically $+\infty$. Let (x, y) with $y \in s(x)$. We shall prove that $y \in \partial f(x)$ for which it is enough to show that for each $\alpha < f(x)$ we have the following inequality:

$$f(z) > \alpha + (z - x) \cdot y, \qquad \forall z \in \mathbb{R}^n.$$

To prove the last inequality, by the definition of supremum there exist (x_i, y_i) with $y_i \in s(x_i)$, $0 \leq i \leq N$, such that

$$\alpha < (x - x_N) \cdot y_N + (x_N - x_{N-1}) \cdot y_{N-1} + \cdots + (x_2 - x_1) \cdot y_1 + (x_1 - x_0) \cdot y_0,$$

and adding $(z - x) \cdot y$ to this inequality we get

$$\begin{aligned} \alpha + (z - x) \cdot y &< (z - x) \cdot y + (x - x_N) \cdot y_N \\ &\quad + (x_N - x_{N-1}) \cdot y_{N-1} + \cdots + (x_2 - x_1) \cdot y_1 + (x_1 - x_0) \cdot y_0 \\ &\leq f(z) \end{aligned}$$

and we are done.

To prove the second part of the theorem, let $y_i \in \partial f(x_i)$ with $0 \leq i \leq N$. Then

$$f(x) \geq f(x_i) + (x - x_i) \cdot y_i$$

for all x and $0 \leq i \leq N$. Hence

$$f(x_{i+1}) - f(x_i) \geq (x_{i+1} - x_i) \cdot y_i$$

for $0 \leq i \leq N - 1$. Adding these inequalities over $0 \leq i \leq N - 1$ yields

$$f(x_N) - f(x_0) \geq \sum_{i=0}^{N-1} (x_{i+1} - x_i) \cdot y_i,$$

and adding $(x_0 - x_N) \cdot y_N$ yields

$$f(x_N) - f(x_0) + (x_0 - x_N) \cdot y_N \geq \sum_{i=0}^{N-1} (x_{i+1} - x_i) \cdot y_i + (x_0 - x_N) \cdot y_N.$$

But $0 \geq f(x_N) - f(x_0) + (x_0 - x_N) \cdot y_N$, since $y_N \in \partial f(x_N)$, and the conclusion follows. □

The following lemma shows that the sub differential determines the function up to a constant.

Lemma 8.3 *Let* $\Omega \subset \mathbb{R}^n$ *be convex and let* $f, g : \Omega \to \mathbb{R}$ *be convex functions. If* $\partial f(x) \cap \partial g(x) \neq \emptyset$ *for all* $x \in \Omega$*, then* $f(x) = g(x) + C$ *for all* $x \in \Omega$ *and some constant* C*.*

Proof Let $x, y \in \Omega$, and let $x^* \in \partial f(x) \cap \partial g(x)$ and $y^* \in \partial f(y) \cap \partial g(y)$. For $N \geq 1$ integer, let

$$x_k = \frac{k}{N} y + \left(1 - \frac{k}{N}\right) x, \qquad 0 \leq k \leq N.$$

Pick $x_k^* \in \partial f(x_k) \cap \partial g(x_k)$ for $1 \leq k \leq N-1$, and set $x_0^* = x^*$ and $x_N^* = y^*$. Obviously, $x_0 = x$ and $x_N = y$, and so $x_j^* \in \partial f(x_j) \cap \partial g(x_j)$ for $0 \leq j \leq N$. We have

$$f(x_{k+1}) - f(x_k) \geq (x_{k+1} - x_k) \cdot x_k^*, \qquad 0 \leq k \leq N-1,$$

and,

$$g(x_k) - g(x_{k+1}) \geq (x_k - x_{k+1}) \cdot x_{k+1}^*, \qquad 0 \leq k \leq N-1.$$

Adding together these inequalities over $0 \leq k \leq N-1$ yields

$$\begin{aligned} f(y) - f(x) + g(x) - g(y) &\geq \frac{1}{N} \sum_{k=0}^{N-1} (y - x) \cdot x_k^* - \frac{1}{N} \sum_{k=0}^{N-1} (y - x) \cdot x_{k+1}^* \\ &= \frac{1}{N} (y - x) \cdot (x_0^* - x_N^*) \\ &= \frac{1}{N} (y - x) \cdot (x^* - y^*) \to 0, \end{aligned}$$

as $N \to \infty$. So $f(y) - g(y) \geq f(x) - g(x)$, for all $x, y \in \Omega$. Reversing the roles of x and y we obtain the lemma. □

Corollary 8.4 *If* f, g *are convex functions in* $\mathbb{R}^n$ *as in Theorem 8.2, i.e.,* s *is cyclically monotone with* $s \subset \partial f$, $s \subset \partial g$*, and* $s(x) \neq \emptyset$ *for all* $x \in \mathbb{R}^n$*, then* $f = g + C$*.*

Remark 8.5 Another view of the above is the following. If f, g are convex and $\partial f(x) \subset \partial g(x)$ for a.e. x, then $f = g + C$. In fact, by Rademacher theorem f and g are differentiable a.e., so $\partial f(x) = Df(x)$ and $\partial g(x) = Dg(x)$ a.e. and therefore $Df(x) = Dg(x)$ a.e. Hence the weak gradient of $f - g$ is zero, and since f, g are continuous the conclusion follows.

Chapter 9
Quadratic Cost

Abstract It is shown that optimal maps for the quadratic cost are cyclically monotone and the corresponding pde is calculated.

9.1 Cyclical Monotonicity of the Optimal Map: Heuristics

Let μ, μ^* be two finite measures in $\mathbb{R}^n$ with compact support and let $s : \mathbb{R}^n \to \mathbb{R}^n$ be measure preserving μ to μ^*, i.e., $\mu(s^{-1}(E)) = \mu^*(E)$ for each Borel set $E \subset \mathbb{R}^n$, and minimizing the integral

$$I(s) = \int_{\mathbb{R}^n} |x - s(x)|^2 \, d\mu(x),$$

among all measure preserving maps μ to μ^*. We show that s is cyclically monotone.

Fix $x_1, \cdots, x_N$ arbitrary distinct points in the support of μ, and disjoint balls $B_{r_k}(x_k)$ with r_k to be chosen as follows. Let $m = \min_{1 \le i,j \le N} |x_i - x_j|$ so the balls $B_{m/3}(x_k)$ are all disjoint. If *μ is absolutely continuous*, then $\mu(B_\delta(x_k)) \to 0$ as $\delta \to 0$. Let $M = \min_{1 \le k \le N} \mu(B_{m/3}(x_k))$, and let $0 < \epsilon < M$. By continuity of $\mu(B_\delta(x_k))$ in δ, for each k pick $0 < r_k \le m/3$ such that $\mu(B_{r_k}(x_k)) = \epsilon$.

Let

$$F_k = s(B_{r_k}(x_k)) \qquad 1 \le k \le N.$$

Assuming s is 1-to-1 we have

$$s^{-1}(F_k) = B_{r_k}(x_k),$$

and since s is measure preserving and the choice of r_k

$$\mu(s^{-1}(F_k)) = \mu^*(F_k) = \epsilon.$$

C. E. Gutiérrez, *Optimal Transport and Applications to Geometric Optics*, SpringerBriefs on PDEs and Data Science,
https://doi.org/10.1007/978-981-99-4867-3_9

Let $\sigma : \{1, \cdots, N\} \to \{1, \cdots, N\}$ be a permutation. Suppose is possible to construct a measure preserving map s_σ such that

$$s_\sigma(B_{r_k}(x_k)) = F_{\sigma(k)} \qquad 1 \le k \le N$$
$$s_{\mathbf{s}}(x_k) = x_{\mathbf{s}(k)} \qquad 1 \le k \le N$$
$$s_\sigma = s \qquad \text{on the set } \mathbb{R}^n \setminus \cup_1^N B_{r_k}(x_k).$$

Since s is a minimizer we have $I(s) \le I(s_\sigma)$ so expanding the dot product yields

$$\int_{\mathbb{R}^n} \left(|x|^2 - 2x \cdot s(x) + |s(x)|^2\right) d\mu(x)$$
$$\le \int_{\mathbb{R}^n} \left(|x|^2 - 2x \cdot s_\sigma(x) + |s_\sigma(x)|^2\right) d\mu(x). \tag{9.1}$$

Since μ is finite and compactly supported we have $\int_{\mathbb{R}^n} |x|^2 \, d\mu(x) < \infty$ and since s, s_σ are both measure preserving from (5.1) we have

$$\int_{\mathbb{R}^n} |s(x)|^2 \, d\mu(x) = \int_{\mathbb{R}^n} |y|^2 \, d\mu^*(y) = \int_{\mathbb{R}^n} |s_\sigma(x)|^2 \, d\mu(x).$$

So from (9.1) we obtain

$$\int_{\mathbb{R}^n} x \cdot (s(x) - s_\sigma(x)) \, d\mu(x) \ge 0,$$

and since $s = s_\sigma$ on $\mathbb{R}^n \setminus \cup_1^N B_{r_k}(x_k)$ we obtain

$$\sum_{k=1}^N \int_{B_{r_k}(x_k)} x \cdot (s(x) - s_\sigma(x)) \, d\mu(x) \ge 0.$$

Dividing this expression by $\mu(B_{r_k}(x_k)) = \epsilon$ yields

$$\sum_{k=1}^N \fint_{B_{r_k}(x_k)} x \cdot (s(x) - s_\sigma(x)) \, d\mu(x) \ge 0.$$

Letting $\epsilon \to 0$ we get that $r_k \to 0$ and then using the differentiation theorem for integrals, assuming all maps and measures are well behaved, we obtain

$$\sum_{k=1}^N x_k \cdot (s(x_k) - s_\sigma(x_k)) \ge 0.$$

If we denote $X = (x_1, \cdots, x_N)$, $s(X) = (s(x_1), \cdots, s(x_N))$, and $s_{\mathbf{s}}(X) = (s(x_{\mathbf{s}(1)}), \cdots, s(x_{\mathbf{s}(N)}))$, then

$$X \cdot s(X) \geq X \cdot s_{\mathbf{s}}(X),$$

or equivalently

$$|X - s(X)|^2 \leq |X - s_{\mathbf{s}}(X)|^2.$$

Since the points $x_1, \cdots, x_N$ are arbitrary and σ is arbitrary, this means that the map s is cyclically monotone.

9.2 Pde for the Quadratic Cost

Suppose we have the optimal map minimizing $\int_{\mathbb{R}^n} |x - s(x)|^2 \, d\mu(x)$ over all measure preserving maps s from μ to μ^*, both measures with compact support. Let us denote by s the optimal map. From Lemma 6.6, $s = \partial u$ a.e. with u convex in $\mathbb{R}^n$.[1] From the measure preserving Eq. (5.1) and assuming all the smoothness needed from the map, we have that

$$\int_{\mathbb{R}^n} v(Du(x)) \, d\mu(x) = \int_{\mathbb{R}^n} v(y) \, d\mu^*(y)$$

for each v continuous. Suppose in addition that $d\mu = g(x)\, dx$ and $d\mu^* = h(y)\, dy$. So we have the equation

$$\int_{\mathbb{R}^n} v(Du(x)) \, g(x) \, dx = \int_{\mathbb{R}^n} v(y) \, h(y) \, dy.$$

Assuming all the differentiability needed, making the change of variables $y = Du(x)$ in the first integral yields

$$\begin{aligned} \int_{\mathbb{R}^n} v(Du(x)) \, g(x) \, dx &= \int_{\mathbb{R}^n} v(y) \, g((Du)^{-1}(y)) \, \frac{1}{\det(D^2u)((Du)^{-1}(y))} \, dy \\ &= \int_{\mathbb{R}^n} v(y) \, h(y) \, dy, \end{aligned}$$

[1] The optimal map is the subdifferential of the function $-(\phi(x) - |x|^2)/2$ where ϕ is $|x - y|^2$-concave.

for all v continuous and therefore $g((Du)^{-1}(y))\,\dfrac{1}{\det(D^2u)((Du)^{-1}(y))} = h(y)$. Returning to the old variables, we obtain that u satisfies the Monge-Ampère type equation

$$\det D^2u(x) = \frac{g(x)}{h(Du(x))}.$$

Chapter 10
Brenier's Polar Factorization Theorem

Abstract A detailed proof of Brenier's theorem is given.

Theorem 10.1 *Let Ω be a bounded domain in $\mathbb{R}^n$ and $\mathcal{F} : \Omega \to \mathbb{R}^n$ be a field such that $\mathcal{F}^{-1}(E)$ is Lebesgue measurable for each E Borel subset of $\mathbb{R}^n$, $|\mathcal{F}(x)| \leq M$ for all $x \in \Omega$, and satisfies $|\mathcal{F}^{-1}(E)| = 0$ for each Borel set E with $|E| = 0$.*

Then there is a convex function u in $\mathbb{R}^n$, and a measure preserving map $s : \Omega \to \Omega$, i.e., $|s^{-1}(E)| = |E|$ for each Borel set $E \subset \Omega$, such that

$$\mathcal{F}(x) = Du(s(x)) \qquad \text{for a.e. } x \in \Omega.$$

The map s and the function u are unique a.e.

Proof We have $\mathcal{F}(\Omega) \subset B_M(0)$, and define

$$\mu(E) = |\mathcal{F}^{-1}(E)|$$

for each Borel set $E \subset B_M(0)$; μ is a finite Borel measure in $B_M(0)$.[1] Since $\mu(E) = 0$ for each E with Lebesgue measure zero, it follows that μ is absolutely continuous with respect to Lebesgue measure and so there is a function $f \in L^1(B_M(0))$ such that $\mu = f\,dx$. We also have

$$\mu\left(B_M(0)\right) = |F^{-1}(B_M(0))| = |\Omega|.$$

We apply the optimal transport theory to the measure spaces $(B_M(0), \mu)$ and (Ω, dx) with the quadratic cost $c(x, m) = |x - m|^2$, $x \in B_M(0)$ and $m \in \Omega$. Let T be the optimal map minimizing $\int_{B_M(0)} |x - T(x)|^2\, d\mu(x)$ among all maps from $B_M(0)$ to Ω that preserve the measures μ to dx. That is,

$$\mu\left(T^{-1}(E)\right) = |E| \qquad \forall E \subset \Omega$$

[1] The σ-additivity follows because if $E \cap F = \emptyset$, then $F^{-1}(E) \cap F^{-1}(F) = \emptyset$.

C. E. Gutiérrez, *Optimal Transport and Applications to Geometric Optics*,
SpringerBriefs on PDEs and Data Science,
https://doi.org/10.1007/978-981-99-4867-3_10

Borel sets. From Lemma 6.6, there is $\psi : B_M(0) \to \Omega$ c-concave such that $\mathcal{N}_{c,\psi} = T$ a.e. If we let $\phi(x) := -\dfrac{\psi(x) - |x|^2}{2}$, ϕ is convex and it is easy to see that $\mathcal{N}_{c,\psi}(x) = \partial\phi(x)$ for each $x \in B_M(0)$, where ∂ is the standard sub differential.

Let ϕ^* be the Legendre transform of ϕ: $\phi^*(y) = \sup_{x \in B_M(0)} (x \cdot y - \phi(x))$; $y \in \mathbb{R}^n$. Since ϕ^* is convex, it is differentiable a.e. and $\nabla\phi^*$ denotes the gradient. Since ϕ is convex, the gradient $\nabla\phi(x)$ exists for a.e. $x \in B_M(0)$. We claim that

$$\nabla\phi^*\left(\nabla\phi(x)\right) = x, \quad \text{for a.e. } x \in B_M(0). \tag{10.1}$$

This follows from the argument at the beginning of Sect. 7. Indeed, let $G = \{x \in B_M(0) : \phi \text{ is differentiable at } x\}$, $M = \{y \in \partial\phi\left(B_M(0)\right) : \phi^* \text{ is differentiable at } y\}$, and $F = (\partial\phi)^{-1}(M) \cap G$. We have $\partial\phi(F) = M$, $\nabla\phi : F \to M$ is bijective and $\left|(\partial\phi)^{-1}(\Omega \setminus M)\right| = 0$. Set $E_1 = (\partial\phi)^{-1}(\Omega \setminus M)$. Then $B_M(0) = E_1 \cup (\partial\phi)^{-1}(M)$ with $|E_1| = 0$. Also $B_M(0) = G \cup E_2$ with $|E_2| = 0$, since ϕ is convex. Then $F = B_M(0) \setminus (E_1 \cup E_2)$, and so almost all points in $B_M(0)$ belong to F. Now if $x \in F$, then $\nabla\phi(x) \in M$ and so ϕ^* is differentiable at $\nabla\phi(x)$ for all $x \in F$. On the other hand, $p \in \partial\phi(x)$ implies $x \in \partial\phi^*(p)$ (see Exercise 12(b)), and therefore $x \in \partial\phi^*\left(\nabla\phi(x)\right)$ for all $x \in F$ and so (10.1) follows for all $x \in F$.

Let us now define $s : \Omega \to \Omega$ by

$$s(x) = \partial\phi\left(\mathcal{F}(x)\right).$$

We first observe that s is well defined for all $x \in \Omega$ as a multivalued map. Since $\Omega = \mathcal{F}^{-1}\left(B_M(0)\right) = \mathcal{F}^{-1}\left(B_M(0) \setminus F\right) \cup \mathcal{F}^{-1}(F)$ and $\left|\mathcal{F}^{-1}\left(B_M(0) \setminus F\right)\right| = 0$, it follows that almost all points in Ω belong to $\mathcal{F}^{-1}(F)$. Consequently,

$$s(x) = \nabla\phi\left(\mathcal{F}(x)\right), \quad \text{for all } x \in \mathcal{F}^{-1}(F)$$

that is, $s(x)$ is single valued a.e. Applying $\nabla\phi^*$ on both sides we get from (10.1) the desired representation formula

$$\mathcal{F}(x) = \nabla\phi^*\left(s(x)\right), \quad \text{for all } x \in F.$$

It remains to show that s is measure preserving μ to dx. In fact, let $E \subset \Omega$ be a Borel set. Since $T = \partial\phi$ preserves μ to dx we obtain

$$|s^{-1}(E)| = \left|F^{-1}\left((D\phi)^{-1}(E)\right)\right| = \mu\left((D\phi)^{-1}(E)\right) = \mu\left(T^{-1}(E)\right) = |E|.$$

Therefore the function $u = \phi^*$ yields the desired decomposition. □

Chapter 11
Benamou and Brenier Formula

Abstract A proof of a dynamic formulation of optimal transport due to Benamou and Brenier is included in this chapter.

We shall prove the following theorem.

Theorem 11.1 ([1]) *Let* $1 \leq p < \infty$ *and let* ρ_i*,* $i = 0, 1$*, be smooth probability densities with compact support in* $\mathbb{R}^n$*. Then*

$$W_p(\rho_0, \rho_1)^p = \inf \left\{ T^{p-1} \int_{\mathbb{R}^n} \int_0^T \rho(x,t)\, |v(x,t)|^p \, dx\, dt \right\} \tag{11.1}$$

where the infimum is over all $\rho = \rho(x,t) \geq 0$ *and smooth bounded* $v(x,t) \in \mathbb{R}^n$ *satisfying the continuity equation*

$$\rho_t + div\,(\rho\, v) = 0, \qquad \rho(x,0) = \rho_0(x), \quad \rho(x,T) = \rho_1(x). \tag{11.2}$$

Proof Let $v(x,t) \in \mathbb{R}^n$ be bounded smooth and let $\sigma(x,t)$ solve

$$\sigma_t(x,t) = v(\sigma(x,t),t), \qquad \sigma(x,0) = x, \tag{11.3}$$

and let $\rho(x,t)$ satisfy (11.2). Then for each $f = f(x,t)$ smooth we have

$$\int_{\mathbb{R}^n} \int_0^T f(y,t)\, \rho(y,t)\, dy\, dt = \int_{\mathbb{R}^n} \int_0^T f(\sigma(x,t),t)\, \rho(x,0)\, dx\, dt. \tag{11.4}$$

C. E. Gutiérrez, *Optimal Transport and Applications to Geometric Optics*, SpringerBriefs on PDEs and Data Science,
https://doi.org/10.1007/978-981-99-4867-3_11

This formula follows because of the following. First, (11.3) implies that $\frac{\partial}{\partial t} J(x,t) = (\operatorname{div} v(\sigma(x,t),t))\, J(x,t)$ where $J(x,t) = \det(D_x\sigma(x,t))$ is the Jacobian. Second, this in turn implies that

$$\frac{d}{dt}\int_{\sigma(\Omega,t)} f(x,t)\,\rho(x,t)\,dx = \int_{\sigma(\Omega,t)} \rho(x,t)\,\left(f_t(x,t) + \nabla_x f(x,t)\cdot v(x,t)\right)\,dx$$

for each f smooth, each ρ satisfying the continuity Eq. (11.2), and each domain Ω. In particular, if $f = 1$, we get $\int_{\sigma(\Omega,t)} \rho(x,t)\,dx = \int_\Omega \rho(x,0)\,dx$ since $\sigma(x,0) = x$. Changing variables and shrinking Ω we then get $\rho(\sigma(x,t),t)\,|J(x,t)| = \rho(x,0)$ for each x and so (11.4) follows.

Our first goal is to show that the mapping $s(x) := \sigma(x,T)$ is measure preserving ρ_0 to ρ_1. By approximation, (11.4) holds when $f(x,t) = g(x)\,\chi_{(a,T)}(t)$ with g smooth. So

$$\begin{aligned}\int_{\mathbb{R}^n}\int_a^T g(y)\,\rho(y,t)\,dy\,dt &= \int_{\mathbb{R}^n}\int_a^T g(\sigma(x,t))\,\rho(x,0)\,dx\,dt \\ &= \int_a^T \left(\int_{\mathbb{R}^n} g(\sigma(x,t))\,\rho_0(x)\,dx\right)dt.\end{aligned} \tag{11.5}$$

Next write

$$\begin{aligned}&\int_{\mathbb{R}^n} g(\sigma(x,t))\,\rho_0(x)\,dx \\ &= \int_{\mathbb{R}^n} (g(\sigma(x,t)) - g(\sigma(x,T))\,\rho_0(x)\,dx + \int_{\mathbb{R}^n} g(\sigma(x,T))\,\rho_0(x)\,dx \\ &= A(t) + \int_{\mathbb{R}^n} g(\sigma(x,T))\,\rho_0(x)\,dx.\end{aligned}$$

Now for $t \le T$

$$\begin{aligned}|A(t)| &\le \int_{\mathbb{R}^n} |g(\sigma(x,t)) - g(\sigma(x,T)|\,\rho_0(x)\,dx \\ &\le \|\nabla g\|_\infty\,\|\rho_0\|_\infty \int_B |\sigma(x,t) - \sigma(x,T)|\,dx \\ &\qquad \text{since } \rho_0 \text{ is bounded with compact support } B \\ &= \|\nabla g\|_\infty\,\|\rho_0\|_\infty \int_B \left|\int_t^T \sigma_t(x,s)\,ds\right|\,dx \\ &= \|\nabla g\|_\infty\,\|\rho_0\|_\infty \int_B \left|\int_t^T v\,((x,s),s)\,ds\right|\,dx \quad \text{from (11.3)} \\ &\le \|\nabla g\|_\infty\,\|\rho_0\|_\infty\, M\,|B|\,(T-t) := C\,(T-t) \\ &\qquad \text{assuming } v \text{ is bounded by } M.\end{aligned}$$

From (11.5) we then get

$$\int_{\mathbb{R}^n}\int_a^T g(y)\,\rho(y,t)\,dy\,dt = \int_a^T A(t)\,dt + (T-a)\int_{\mathbb{R}^n} g(\sigma(x,T))\,\rho_0(x)\,dx. \tag{11.6}$$

Since

$$\left|\int_a^T A(t)\,dt\right| \le C\,(T-a)^2,$$

dividing (11.6) by $T-a$ and letting $a \to T^-$ yields

$$\int_{\mathbb{R}^n} g(y)\,\rho(y,T)\,dy = \int_{\mathbb{R}^n} g(\sigma(x,T))\,\rho_0(x)\,dx,$$

for each g smooth in $\mathbb{R}^n$. Since $\rho(y,T) = \rho_1(x)$, we then get that the map $s(x) := \sigma(x,T)$ preserves the measures $\rho_i(x)\,dx$, $i = 0, 1$.

We now write

$$\begin{aligned}
&\int_{\mathbb{R}^n}\int_0^T \rho(x,t)\,|v(x,t)|^p\,dx\,dt \\
&= \int_{\mathbb{R}^n}\int_0^T \rho_0(x)\,|v(\sigma(x,t),t)|^p\,dx\,dt \quad \text{from (11.4)} \\
&= \int_{\mathbb{R}^n}\int_0^T \rho_0(x)\,|\sigma_t(x,t)|^p\,dx\,dt \quad \text{from (11.3)} \\
&= \int_{\mathbb{R}^n} \rho_0(x)\left(\int_0^T |\sigma_t(x,t)|^p\,dt\right)dx.
\end{aligned}$$

By Hölder's inequality

$$\begin{aligned}
⨍_0^T |\sigma_t(x,t)|^p\,dt &\ge \left(⨍_0^T |\sigma_t(x,t)|\,dt\right)^p \ge \left|⨍_0^T \sigma_t(x,t)\,dt\right|^p \\
&= \frac{1}{T^p}\,|\sigma(x,T) - \sigma(x,0)|^p.
\end{aligned}$$

Therefore

$$\begin{aligned}
T^{p-1}\int_{\mathbb{R}^n}\int_0^T \rho(x,t)\,|v(x,t)|^p\,dx\,dt &\ge \int_{\mathbb{R}^n} \rho_0(x)\;|\sigma(x,T) - \sigma(x,0)|^p\,dx \\
&= \int_{\mathbb{R}^n} |s(x) - x|^p\;\rho_0(x)\,dx.
\end{aligned}$$

If s_0 is a measure preserving mapping minimizing the Monge problem, we then obtain

$$T^{p-1} \int_{\mathbb{R}^n} \int_0^T \rho(x,t)\,|v(x,t)|^p \,dx\,dt \geq \int_{\mathbb{R}^n} |s(x)-x|^p \,\rho_0(x)\,dx$$
$$\geq \int_{\mathbb{R}^n} |s_0(x)-x|^p \,\rho_0(x)\,dx, \qquad (11.7)$$

for all ρ and v satisfying (11.2). Therefore the infimum in (11.1) is greater than or equal to $W_p(\rho_0, \rho_1)^p$.

To show the opposite inequality, we construct a field v so that the left and the right hand sides of (11.7) are equal. Let us define

$$\sigma_0(x,t) = x + \frac{t}{T}(s_0(x) - x). \qquad (11.8)$$

Suppose this mapping is invertible (see Remark 11.2) and assume all the smoothness needed, i.e., if $z = x + \frac{t}{T}(s_0(x) - x)$, let

$$x = \Phi(z,t).$$

So $\Phi\,(\sigma_0(x,t), t) = x$. Define

$$v(z,t) = \frac{s_0\,(\Phi(z,t)) - \Phi(z,t)}{T}.$$

Then

$$(\sigma_0)_t(x,t) = \frac{s_0(x) - x}{T} = \frac{s_0\,(\Phi\,(\sigma_0(x,t),t)) - \Phi\,(\sigma_0(x,t),t)}{T} = v\,(\sigma_0(x,t),t)$$

and $\sigma_0(x,0) = x$; $v(p,t) = \dfrac{s_0\,(\Phi(p,t)) - \Phi(p,t)}{T}$. Next let ρ solve the problem (11.2) with the v just defined. We show that for these ρ and v we have

$$T^{p-1} \int_{\mathbb{R}^n} \int_0^T \rho(x,t)\,|v(x,t)|^p \,dx\,dt = \int_{\mathbb{R}^n} |s_0(x)-x|^p \,\rho_0(x)\,dx.$$

Indeed, since (11.4) holds for this ρ and v, for each $f(x,t)$ smooth with $\sigma = \sigma_0$, we can apply (11.4) with $f(x,t) = |v(x,t)|^p$ to obtain

$$
\begin{aligned}
&T^{p-1} \int_{\mathbb{R}^n} \int_0^T \rho(x,t)\, |v(x,t)|^p \, dx\, dt \\
&= T^{p-1} \int_{\mathbb{R}^n} \int_0^T \rho_0(x)\, |v(\sigma_0(x,t),t)|^p \, dx\, dt \\
&= T^{p-1} \int_{\mathbb{R}^n} \int_0^T \rho_0(x)\, |(\sigma_0)_t(x,t)|^p \, dx\, dt \\
&= T^{p-1} \int_{\mathbb{R}^n} \rho_0(x) \left(\int_0^T \left| \frac{s_0(x) - x}{T} \right|^p dt \right) dx \\
&= \int_{\mathbb{R}^n} |s_0(x) - x|^p \, \rho_0(x)\, dx
\end{aligned}
$$

as desired. So the proof is complete. □

Remark 11.2 Suppose $\phi : \mathbb{R}^n \to \mathbb{R}_{\geq 0}$ is smooth, even, and convex, and consider the cost $c(x,y) = \phi\left((x-y)/T\right)$. We can easily adapt the proof above to show that

$$
\begin{aligned}
&\min \left\{ \int_{\mathbb{R}^n} c\left(s(x), x\right) \rho_0(x)\, dx : s \text{ measure preserving } \rho_0 \text{ to } \rho_1 \right\} \\
&= \min \left\{ \frac{1}{T} \int_{\mathbb{R}^n} \int_0^T \rho(x,t)\, \phi\left(v(x,t)\right)\, dx\, dt \right\}.
\end{aligned}
$$

In fact, by Jensen's inequality

$$
\begin{aligned}
c\left(\sigma(x,T), \sigma(x,0)\right) = \phi\left(\frac{\sigma(x,T) - \sigma(x,0)}{T} \right) &= \phi\left(\fint_0^T \sigma_t(x,s)\, ds \right) \\
&\leq \fint_0^T \phi\left(\sigma_t(x,s)\right)\, ds = \fint_0^T \phi\left(v\left(\sigma(x,s), s\right)\right)\, ds
\end{aligned}
$$

by (11.3). Hence with $s(x) = \sigma(x,T)$

$$
\begin{aligned}
\int_{\mathbb{R}^n} c\left(s(x), x\right)\, \rho_0(x)\, dx &\leq \frac{1}{T} \int_{\mathbb{R}^n} \int_0^T \phi\left(v\left(\sigma(x,s), s\right)\right)\, \rho_0(x)\, dx\, ds \\
&= \frac{1}{T} \int_{\mathbb{R}^n} \int_0^T \phi\left(v(y,t)\right)\, \rho(y,t)\, dy\, dt
\end{aligned}
$$

from (11.4).

To show equality we choose σ_0 as in (11.8), v as in the proof above, and let $\rho(x,t)$ be the solution to (11.2) with this v. Then we can apply (11.4) with these ρ and σ_0 and with $f(x,t) = \phi\,(v(x,t))$ to obtain

$$\begin{aligned}\frac{1}{T}\int_{\mathbb{R}^n}\int_0^T \rho(x,t)\,\phi\,(v(x,t))\,dx\,dt &= \frac{1}{T}\int_{\mathbb{R}^n}\int_0^T \rho_0(x)\,\phi\,(v(\sigma_0(x,t),t))\,dx\,dt\\ &= \frac{1}{T}\int_{\mathbb{R}^n}\int_0^T \rho_0(x)\,\phi\,((\sigma_0)_t(x,t))\,dx\,dt\\ &= \frac{1}{T}\int_{\mathbb{R}^n}\rho_0(x)\left(\int_0^T \phi\left(\frac{s_0(x)-x}{T}\right)dt\right)dx\\ &= \int_{\mathbb{R}^n} c\,(s_0(x),x)\;\rho_0(x)\,dx.\end{aligned}$$

Remark 11.3 Suppose $c(x,y) = \phi(x-y)$ is the cost in Remark 11.2 with $T = 1$. We show the invertibility of the mapping defined in (11.8). From Lemmas 6.6 and 6.7 the optimal map s_0 is the c-subdifferential of a c-concave function ψ, i.e., $s_0(x) = \mathcal{N}_{c,\psi}(x)$ for ρ_0-a.e. x, i.e., there is $E \subset \text{supp}\,(\rho_0)$ with $|\text{supp}\,(\rho_0) \setminus E| = 0$ such that $s_0(x)$ is single valued for all $x \in E$. We also have that $\mathcal{N}_{c,\psi}$ is c-cyclically monotone with obvious changes in Definition 8.1. We prove that $(1-t)x + t\,s_0(x)$ is injective in E, for each $0 < t < 1$. Let $x, y \in E$ with $(1-t)x + t\,s_0(x) = (1-t)y + t\,s_0(y) = z$. Then

$$\begin{aligned} t\,(s_0(y)-y) + (1-t)\,(s_0(x)-x) &= s_0(x) - y\\ t\,(s_0(x)-x) + (1-t)\,(s_0(y)-y) &= s_0(y) - x. \end{aligned}$$

On the other hand, since from its definition $\mathcal{N}_{c,\psi}$ is c-cyclically monotone:

$$c(x,s_0(x)) + c(y,s_0(y)) \le c(x,s_0(y)) + c(y,s_0(x)), \quad x, y \in E,$$

it follows that

$$\begin{aligned}\phi\,(x - s_0(x)) + \phi\,(y - s_0(y)) &\le \phi\,(x - s_0(y)) + \phi\,(y - s_0(x))\\ &= \phi\,(t\,(s_0(x)-x) + (1-t)\,(s_0(y)-y))\\ &\quad + \phi\,(t\,(s_0(y)-y) + (1-t)\,(s_0(x)-x))\\ &\le \phi\,(x - s_0(x)) + \phi\,(y - s_0(y)).\end{aligned}$$

Since $0 < t < 1$ and ϕ is strictly convex, we get $s_0(x) - x = s_0(y) - y$, and therefore $x = y$.

Chapter 12
Snell's Law of Refraction

Abstract The Snell law in vector form is stated and proved using wavefronts. Surfaces for uniform refraction both in the near field and far field cases are calculated.

12.1 In Vector Form

Suppose Γ is a surface in $\mathbb{R}^3$ that separates two media I and II that are homogeneous and isotropic. Let v_1 and v_2 be the velocities of propagation of light in the media I and II respectively. The index of refraction of the medium I is by definition $n_1 = c/v_1$, where c is the velocity of propagation of light in the vacuum, and similarly $n_2 = c/v_2$. If a ray of light[1] having direction $x \in S^2$ and traveling through the medium I hits Γ at the point P, then this ray is refracted in the direction $m \in S^2$ through the medium II according with the Snell law in vector form:

$$n_1(x \times \nu) = n_2(m \times \nu), \tag{12.1}$$

where ν is the unit normal to the surface to Γ at P going towards the medium II. It is assumed here that $x \cdot \nu \geq 0$. The derivation of this formula is in Sect. 12.2. This has several consequences:

(a) the vectors x, m, ν are all on the same plane (called plane of incidence);
(b) the well known Snell law in scalar form

$$n_1 \sin\theta_1 = n_2 \sin\theta_2,$$

where θ_1 is the angle between x and ν (the angle of incidence), θ_2 the angle between m and ν (the angle of refraction),

[1] Since the refraction angle depends on the frequency of the radiation, we assume radiation is monochromatic.

C. E. Gutiérrez, *Optimal Transport and Applications to Geometric Optics*,
SpringerBriefs on PDEs and Data Science,
https://doi.org/10.1007/978-981-99-4867-3_12

From (12.1), $(n_1 x - n_2 m) \times \nu = 0$, which means that the vector $n_1 x - n_2 m$ is parallel to the normal vector ν. If we set $\kappa = n_2/n_1$, then

$$x - \kappa\, m = \lambda \nu, \tag{12.2}$$

for some $\lambda \in \mathbb{R}$. Taking dot products we get $\lambda = \cos\theta_1 - \kappa \cos\theta_2$, $\cos\theta_1 = x \cdot \nu \geq 0$, and $\cos\theta_2 = m \cdot \nu = \sqrt{1 - \sin^2\theta_2} = \sqrt{1 - (n_1/n_2)^2 \sin^2\theta_1} = \sqrt{1 - \kappa^{-2}[1 - (x \cdot \nu)^2]}$, so

$$\lambda = x \cdot \nu - \kappa \sqrt{1 - \kappa^{-2}\left(1 - (x \cdot \nu)^2\right)}. \tag{12.3}$$

Refraction behaves differently for $\kappa < 1$ and for $\kappa > 1$.

12.1.1 $\kappa < 1$

This means $v_1 < v_2$, and so waves propagate in medium II faster than in medium I, or equivalently, medium I is denser than medium II. In this case the refracted rays tend to bent away from the normal, that is the case for example, when medium I is glass and medium II is air. Indeed, from the scalar Snell law, $\sin\theta_1 = \kappa \sin\theta_2 < \sin\theta_2$ and so $\theta_1 < \theta_2$. For this reason, the maximum angle of refraction θ_2 is $\pi/2$ which, from the Snell law in scalar form, is achieved when $\sin\theta_1 = n_2/n_1 = \kappa$. So there cannot be refraction when the incidence angle θ_1 is beyond this critical value, that is, we must have $0 \leq \theta_1 \leq \theta_c = \arcsin\kappa$.[2] Once again from the Snell law in scalar form,

$$\theta_2 - \theta_1 = \arcsin(\kappa^{-1} \sin\theta_1) - \theta_1 \tag{12.4}$$

and it is easy to verify that this quantity is strictly increasing for $\theta_1 \in [0,\ \theta_c]$, and therefore $0 \leq \theta_2 - \theta_1 \leq \dfrac{\pi}{2} - \theta_c$. We then have $x \cdot m = \cos(\theta_2 - \theta_1) \geq \cos(\pi/2 - \theta_c) = \kappa$, and therefore obtain the following physical constraint for refraction:

if $\kappa = n_2/n_1 < 1$ *and a ray of direction* x *through medium I*

is refracted into medium II in the direction m, *then* $m \cdot x \geq \kappa$. (12.5)

Notice also that in this case $\lambda > 0$ in (12.3). Also notice that given x and a normal ν to the interface, a ray with direction x is transmitted to medium II if and only if

$$x \cdot \nu \geq \sqrt{1 - \kappa^2}.$$

[2] If $\theta_1 > \theta_c$, then the phenomenon of total internal reflection occurs, see Fig. 12.1c.

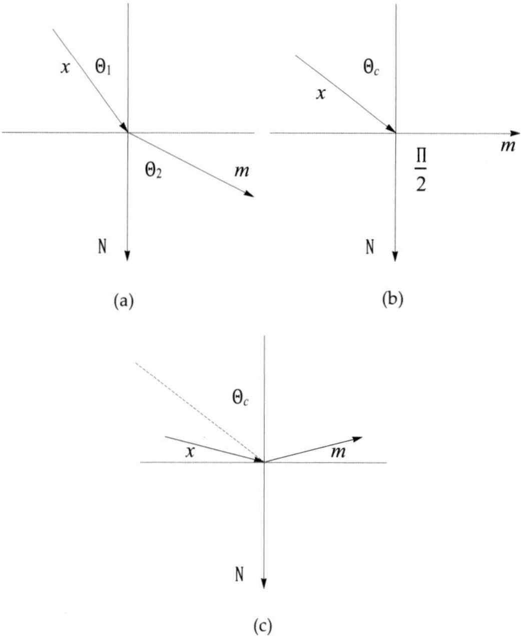

Fig. 12.1 Snell's law $\kappa < 1$, e.g., glass to air. (**a**) $x - \kappa m$ parallel to ν. (**b**) Critical angle. (**c**) Internal reflection

Conversely, given $x, m \in S^2$ with $x \cdot m \geq \kappa$ and $\kappa < 1$, it follows from (12.6) that there exists a hyperplane refracting any ray through medium I with direction x into a ray of direction m in medium II.

12.1.2 $\kappa > 1$

In this case, waves propagate in medium I faster than in medium II, and the refracted rays tend to bent towards the normal. By the Snell law, the maximum angle of refraction denoted by θ_c^* is achieved when $\theta_1 = \pi/2$, and $\theta_c^* = \arcsin(1/\kappa)$. Once again from the Snell law in scalar form

$$\theta_1 - \theta_2 = \arcsin(\kappa \sin \theta_2) - \theta_2 \tag{12.6}$$

which is strictly increasing for $\theta_2 \in [0, \theta_c^*]$, and $0 \leq \theta_1 - \theta_2 \leq \dfrac{\pi}{2} - \theta_c^*$. We therefore obtain the following physical constraint for the case $\kappa > 1$:

if a ray with direction x traveling through medium I

is refracted into a ray in medium II with direction m, then $m \cdot x \geq 1/\kappa$. (12.7)

Notice also that in this case $\lambda < 0$ in (12.3).

On the other hand, by (12.6), if $x, m \in S^2$ with $x \cdot m \geq 1/\kappa$ and $\kappa > 1$, then there exists a hyperplane refracting any ray of direction x through medium I into a ray with direction m in medium II.

We summarize the above discussion on the physical constraints of refraction in the following lemma.

Lemma 12.1 *Let n_1 and n_2 be the indices of refraction of two media I and II, respectively, and $\kappa = n_2/n_1$. Then a light ray in medium I with direction $x \in S^2$ is refracted by some surface into a light ray with direction $m \in S^2$ in medium II if and only if $m \cdot x \geq \kappa$, when $\kappa < 1$; and if and only if $m \cdot x \geq 1/\kappa$, when $\kappa > 1$.*

12.1.3 $\kappa = 1$

This corresponds to reflection. It means

$$x - m = \lambda\, \nu. \tag{12.8}$$

Taking dot products with x and then with m yields $1 - m \cdot x = \lambda\, x \cdot \nu$ and $x \cdot m - 1 = \lambda\, m \cdot \nu$, then $x \cdot \nu = -m \cdot \nu$. Also taking dot product with x in (12.8) then yields $\lambda = 2x \cdot \nu$. Therefore

$$m = x - 2(x \cdot \nu)\nu.$$

12.2 Derivation of the Snell Law

At time t, $\psi(x, y, z, t) = 0$ denotes a surface that separates the part of the space that is at rest with the part of the space that is disturbed by the electric and magnetic fields. For each t fixed, the surface defined by $\psi(x, y, z, t) = 0$ is called a *wave front*. The *light rays* are by definition the orthogonal trajectories to the wave fronts. We assume that $\psi_t \neq 0$ and so we can solve $\psi(x, y, z, t) = 0$ in t obtaining that

$$\phi(x, y, z) = c\,t,$$

where c is the speed of light in vacuum. Therefore, when t runs, we get that the wave fronts are the level sets of the function $\phi(x, y, z)$.

Let us assume that the wave fronts travel in an homogenous and isotropic medium I with refractive index $n_1 (= c/v_1)$, where v_1 is the speed of propagation in medium I. This wave front is transmitted to another homogeneous and isotropic medium II having refractive index n_2. Let Σ be the surface in 3-d separating the media I and II, and suppose it is given by the equations $x = f(\xi, \eta)$, $y = g(\xi, \eta)$ and $z = h(\xi, \eta)$. Let $\phi_1(x, y, z) = c\,t$ be the wave front in medium I and $\phi_2(x, y, z) = c\,t$

be the wave front, to be determined, in medium II. On the surface Σ the two wave fronts agree, that is,

$$\phi_1\left(f(\xi,\eta), g(\xi,\eta), h(\xi,\eta)\right) = \phi_2\left(f(\xi,\eta), g(\xi,\eta), h(\xi,\eta)\right).$$

Differentiating this equation with respect to ξ and η yields

$$\left(\frac{\partial\phi_1}{\partial x} - \frac{\partial\phi_2}{\partial x}\right) f_\xi + \left(\frac{\partial\phi_1}{\partial y} - \frac{\partial\phi_2}{\partial y}\right) g_\xi + \left(\frac{\partial\phi_1}{\partial z} - \frac{\partial\phi_2}{\partial z}\right) h_\xi = 0$$
$$\left(\frac{\partial\phi_1}{\partial x} - \frac{\partial\phi_2}{\partial x}\right) f_\eta + \left(\frac{\partial\phi_1}{\partial y} - \frac{\partial\phi_2}{\partial y}\right) g_\eta + \left(\frac{\partial\phi_1}{\partial z} - \frac{\partial\phi_2}{\partial z}\right) h_\eta = 0.$$

This means that the vector $D\phi_1 - D\phi_2$ is perpendicular to both vectors (f_ξ, g_ξ, h_ξ) and (f_η, g_η, h_η), and therefore it is normal to the surface Σ. Let ν be the outer normal at the surface Σ. Then we have

$$(D\phi_1 - D\phi_2) = \lambda\,\nu, \tag{12.9}$$

for some scalar λ. A light ray $\gamma_1(t)$ in medium I has constant speed v_1 and a light ray $\gamma_2(t)$ in II constant speed v_2. So we have $\phi_1(\gamma_1(t)) = c\,t$ and $\phi_2(\gamma_2(t)) = c\,t$. Differentiating with respect to t yields $D\phi_i(\gamma_i(t)) \cdot \gamma_i'(t) = c$, $i = 1, 2$; $\gamma_i'(t)$ being the direction of propagation of the ray in medium i. Let θ_i be the angle between the vectors $D\phi_i(\gamma_i(t))$, $\gamma_i'(t)$. Since the rays are the orthogonal trajectories to the wave front, we get that $\theta_i = 0$ for $i = 1, 2$, meaning that the vectors $D\phi_i(\gamma_i(t))$, $\gamma_i'(t)$ have the same direction. If γ_i is parametrized so that $|\gamma_i'(t)| = v_i$, we then obtain

$$|D\phi_i(\gamma_i(t))| = \frac{c}{v_i} = n_i.$$

If we let $x = \dfrac{D\phi_1(\gamma_1(t))}{|D\phi_1(\gamma_1(t))|}$ and $m = \dfrac{D\phi_2(\gamma_2(t))}{|D\phi_2(\gamma_2(t))|}$, we then obtain from (12.9) that

$$n_1\,x - n_2\,m = \lambda\,\nu$$

which is equivalent to (12.1).

12.3 Surfaces with the Uniform Refracting Property: Far Field Case

Let $m \in S^2$ be fixed, and we ask the following: if rays of light emanate from the origin inside medium I, what is the surface Γ, interface of the media I and II, that refracts all these rays into rays parallel to m?

Suppose Γ is parameterized by the polar representation $\rho(x)x$ where $\rho > 0$ and $x \in S^2$. Consider a curve on Γ given by $r(t) = \rho(x(t))\,x(t)$ for $x(t) \in S^2$. According to (12.2), the tangent vector $r'(t)$ to Γ satisfies $r'(t) \cdot (x(t) - \kappa\, m) = 0$. That is, $\left([\rho(x(t))]'x(t) + \rho(x(t))x'(t)\right) \cdot (x(t) - \kappa\, m) = 0$, which yields $(\rho(x(t))(1 - \kappa m \cdot x(t)))' = 0$. Therefore

$$\rho(x) = \frac{b}{1 - \kappa\, m \cdot x} \tag{12.10}$$

for $x \in S^2$ and for some $b \in \mathbb{R}$. To understand the surface given by (12.10), we distinguish two cases $\kappa < 1$ and $\kappa > 1$.

12.3.1 Case $\kappa = 1$

When $\kappa = 1$ we see this is a paraboloid. Indeed, let $m = -e_n$, then a point $X = \rho(x)x$ is on the surface (12.10) if $|X| = b - x_n$. The distance from X to the plane $x_n = b$ is $b - x_n$, and the distance from X to 0 is $|X|$. So this is a paraboloid with focus at 0, directrix plane $x_n = b$ and axis in the direction $-e_n$.

12.3.2 Case $\kappa < 1$

Since $\rho(x) > 0$ and $1 - \kappa\, x \cdot m \geq 1 - \kappa > 0$, we have in (12.10) that $b > 0$. We will show that the surface Γ given by (12.10) is an ellipsoid of revolution about the axis of direction m. Suppose for simplicity that $m = e_n$, the nth-coordinate vector. If $y = (y', y_n) \in \mathbb{R}^n$ is a point on Γ, then $y = \rho(x)x$ with $x = y/|y|$. From (12.10), $|y| - \kappa\, y_n = b$, that is, $|y'|^2 + y_n^2 = (\kappa\, y_n + b)^2$ which yields $|y'|^2 + (1 - \kappa^2)y_n^2 - 2\kappa b y_n = b^2$. This surface Γ can be written in the form

$$\frac{|y'|^2}{\left(\dfrac{b}{\sqrt{1-\kappa^2}}\right)^2} + \frac{\left(y_n - \dfrac{\kappa b}{1-\kappa^2}\right)^2}{\left(\dfrac{b}{1-\kappa^2}\right)^2} = 1 \tag{12.11}$$

which is an ellipsoid of revolution about the y_n axis with foci $(0, 0)$ and $(0, 2\kappa b/(1 - \kappa^2))$, major semi-axis= $b/(1 - \kappa^2)$, minor semi-axis= $b/\sqrt{1 - \kappa^2}$, center $(0, \kappa b/(1 - \kappa^2))$, and eccentricity κ. Since $|y| = \kappa y_n + b$ and the physical constraint for refraction (12.5), $\dfrac{y}{|y|} \cdot e_n \geq \kappa$ is equivalent to $y_n \geq \dfrac{\kappa b}{1-\kappa^2}$. That is, for refraction to occur y must be in the upper part of the ellipsoid (12.11); we denote this semi-ellipsoid by $E(e_n, b)$, see Fig. 12.2. To verify that $E(e_n, b)$ has the

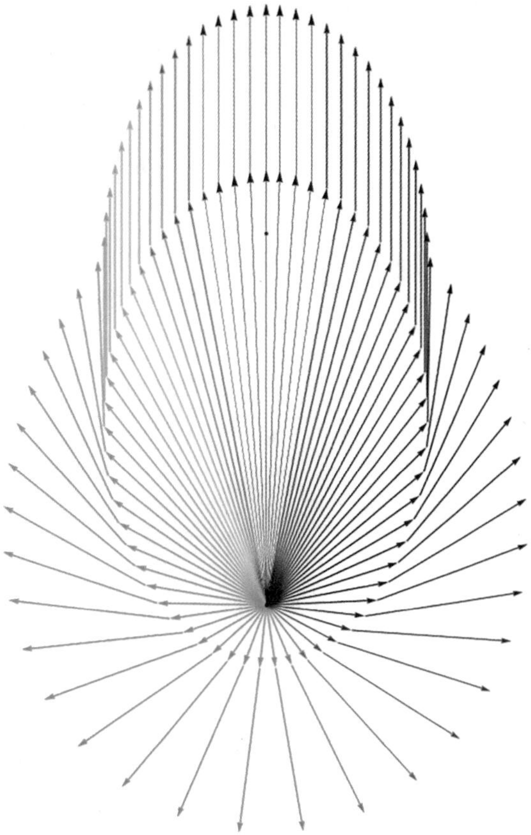

Fig. 12.2 Only half of the ellipse refracts in the direction $m = e_n$

uniform refracting property, that is, it refracts any ray emanating from the origin in the direction e_n, we check that (12.2) holds at each point. Indeed, if $y \in E(e_n, b)$, then $\left(\frac{y}{|y|} - \kappa e_n\right) \cdot \frac{y}{|y|} \geq 1 - \kappa > 0$, and $\left(\frac{y}{|y|} - \kappa e_n\right) \cdot e_n \geq 0$, and so $\frac{y}{|y|} - \kappa e_n$ is an outward normal to $E(e_n, b)$ at y (the normal is calculated from the formula $|y| - \kappa\, y_n = b$).

Rotating the coordinates, it is easy to see that the surface given by (12.10) with $\kappa < 1$ and $b > 0$ is an ellipsoid of revolution about the axis of direction m with foci 0 and $\frac{2\kappa b}{1 - \kappa^2} m$. Moreover, the semi-ellipsoid $E(m, b)$ given by

$$E(m, b) = \{\rho(x)x : \rho(x) = \frac{b}{1 - \kappa\, m \cdot x},\ x \in S^{n-1},\ x \cdot m \geq \kappa\}, \tag{12.12}$$

has the uniform refracting property, any ray emanating from the origin O is refracted in the direction m.

12.3.3 Case $\kappa > 1$

Due to the physical constraint of refraction (12.7), we must have $b < 0$ in (12.10). Define for $b > 0$

$$H(m,b) = \{\rho(x)x : \rho(x) = \frac{b}{\kappa\, m \cdot x - 1},\ x \in S^{n-1},\ x \cdot m \geq 1/\kappa\}. \tag{12.13}$$

We claim that $H(m,b)$ is the sheet with opening in direction m of a hyperboloid of revolution of two sheets about the axis of direction m. To prove the claim, set for simplicity $m = e_n$. If $y = (y', y_n) \in H(e_n, b)$, then $y = \rho(x)x$ with $x = y/|y|$. From (12.13), $\kappa\, y_n - |y| = b$, and therefore $|y'|^2 + y_n^2 = (\kappa\, y_n - b)^2$ which yields $|y'|^2 - (\kappa^2 - 1)\left[\left(y_n - \frac{\kappa b}{\kappa^2 - 1}\right)^2 - \left(\frac{\kappa b}{\kappa^2 - 1}\right)^2\right] = b^2$. Thus, any point y on $H(e_n, b)$ satisfies the equation

$$\frac{\left(y_n - \frac{\kappa b}{\kappa^2 - 1}\right)^2}{\left(\frac{b}{\kappa^2 - 1}\right)^2} - \frac{|y'|^2}{\left(\frac{b}{\sqrt{\kappa^2 - 1}}\right)^2} = 1 \tag{12.14}$$

which represents a hyperboloid of revolution of two sheets about the y_n axis with foci $(0,0)$ and $(0, 2\kappa b/(\kappa^2 - 1))$. Moreover, the upper sheet of this hyperboloid of revolution is given by

$$y_n = \frac{\kappa b}{\kappa^2 - 1} + \frac{b}{\kappa^2 - 1}\sqrt{1 + \frac{|y'|^2}{\left(b/\sqrt{\kappa^2 - 1}\right)^2}}$$

and satisfies $\kappa y_n - b > 0$, and hence has polar equation $\rho(x) = \dfrac{b}{\kappa\, e_n \cdot x - 1}$. Similarly, the lower sheet satisfies $\kappa y_n - b < 0$ and has polar equation $\rho(x) = \dfrac{b}{\kappa\, e_n \cdot x + 1}$. For a general m, by a rotation, we obtain that $H(m,b)$ is the sheet with opening in direction m of a hyperboloid of revolution of two sheets about the axis of direction m with foci $(0,0)$ and $\dfrac{2\kappa b}{\kappa^2 - 1}m$.

Notice that the focus $(0, 0)$ is outside the region enclosed by $H(m, b)$ and the focus $\dfrac{2\kappa b}{\kappa^2 - 1}m$ is inside that region. The vector $\kappa m - \dfrac{y}{|y|}$ is an inward normal to $H(m, b)$ at y, because by (12.13)

$$\left(\kappa m - \frac{y}{|y|}\right) \cdot \left(\frac{2\kappa b}{\kappa^2 - 1}m - y\right) \geq \frac{2\kappa^2 b}{\kappa^2 - 1} - \frac{2\kappa b}{\kappa^2 - 1} - \kappa m \cdot y + |y|$$
$$= \frac{2\kappa b}{\kappa + 1} - b = \frac{b(\kappa - 1)}{\kappa + 1} > 0.$$

Clearly, $\left(\kappa m - \dfrac{y}{|y|}\right) \cdot m \geq \kappa - 1$ and $\left(\kappa m - \dfrac{y}{|y|}\right) \cdot \dfrac{y}{|y|} > 0$. Therefore, $H(m, b)$ satisfies the uniform refraction property.

We remark that one has to use $H(-e_n, b)$ to uniformly refract in the direction $-e_n$, and due to the physical constraint (12.7), the lower sheet of the hyperboloid of Eq. (12.14) cannot refract in the direction $-e_n$.

From the above discussion, we have proved the following.

Lemma 12.2 *Let n_1 and n_2 be the indexes of refraction of two media I and II, respectively, and $\kappa = n_2/n_1$. Assume that the origin O is inside medium I, and $E(m, b)$, $H(m, b)$ are defined by* (12.12) *and* (12.13), *respectively. We have:*

(i) *If $\kappa < 1$ and $E(m, b)$ is the interface of media I and II, then $E(m, b)$ refracts all rays emitted from O into rays in medium II with direction m.*
(ii) *If $\kappa > 1$ and $H(m, b)$ separates media I and II, then $H(m, b)$ refracts all rays emitted from O into rays in medium II with direction m.*

12.4 Uniform Refraction: Near Field Case

The question we now ask is: given a point O inside medium I and a point P inside medium II, find an interface surface $\mathcal{S}$ between media I and II that refracts all rays emanating from the point O into the point P. Suppose O is the origin, and let $X(t)$ be a curve on $\mathcal{S}$. By the Snell law of refraction the tangent vector $X'(t)$ satisfies

$$X'(t) \cdot \left(\frac{X(t)}{|X(t)|} - \kappa \frac{P - X(t)}{|P - X(t)|}\right) = 0.$$

That is,

$$|X(t)|' + \kappa |P - X(t)|' = 0.$$

Therefore $\mathcal{S}$ is the Cartesian oval

$$|X| + \kappa |X - P| = b. \tag{12.15}$$

Since $f(X) = |X| + \kappa|X - P|$ is a convex function, the oval is a convex set.[3]

We need to find and analyze the polar equation of the oval. Write $X = \rho(x)x$ with $x \in S^{n-1}$. Then writing $\kappa|\rho(x)x - P| = b - \rho(x)$, squaring this quantity and solving the quadratic equation yields

$$\rho(x) = \frac{(b - \kappa^2 x \cdot P) \pm \sqrt{(b - \kappa^2 x \cdot P)^2 - (1 - \kappa^2)(b^2 - \kappa^2|P|^2)}}{1 - \kappa^2}. \tag{12.16}$$

Set

$$\Delta(t) = (b - \kappa^2 t)^2 - (1 - \kappa^2)(b^2 - \kappa^2|P|^2). \tag{12.17}$$

12.5 Case $0 < \kappa < 1$

We have

$$\Delta(t) \geq \kappa^2(t - b)^2, \quad \text{for } -|P| \leq t \leq |P|, \text{ with equality if and only if } t = \pm|P|. \tag{12.18}$$

Therefore

$$\Delta(x \cdot P) > \kappa^2(x \cdot P - b)^2, \qquad \text{if } |x \cdot P| < |P|. \tag{12.19}$$

If $b \geq |P|$, then O and P are inside or on the oval, and so the oval cannot refract rays to P. If the oval is non empty, then $\kappa|P| \leq b$. In case $\kappa|P| = b$, the oval reduces to the point O. The only interesting case is then $\kappa|P| < b < |P|$. From the equation of the oval we get that $\rho(x) \leq b$. So we now should decide which values $\pm$ to take in the definition of $\rho(x)$.

Let ρ_+ and ρ_- be the corresponding ρ's. We claim that $\rho_+(x) > b$ and $\rho_-(x) \leq b$. Indeed,

$$\begin{aligned}
\rho_+(x) &= \frac{(b - \kappa^2 x \cdot P) + \sqrt{\Delta(x \cdot P)}}{1 - \kappa^2} \\
&\geq \frac{(b - \kappa^2 x \cdot P) + \kappa|b - x \cdot P|}{1 - \kappa^2} \\
&= b + \frac{\kappa^2(b - x \cdot P) + \kappa|b - x \cdot P|}{1 - \kappa^2} \\
&\geq b.
\end{aligned}$$

---

[3] See the paper [39] where it is mentioned that the demonstration of the optical properties of ovals is given in [44, Book I, Propositions XCVII and XCVIII, pp. 247-248]. See also [34, Chapter VI]. Also the paper [39] contains generalized ovals having more than two foci; see also [40] for more detailed results.

If the equality $\rho_+(x) = b$ holds, then $b = x \cdot P$ and from (12.18) we have $|x \cdot P| = |P|$. So $\rho_+(x) > b$ if $\kappa|P| < b < |P|$. Similarly,

$$\begin{aligned}\rho_-(x) &= \frac{(b - \kappa^2 x \cdot P) - \sqrt{\Delta(x \cdot P)}}{1 - \kappa^2} \\ &\leq \frac{(b - \kappa^2 x \cdot P) - \kappa|b - x \cdot P|}{1 - \kappa^2} \\ &= b + \frac{\kappa^2(b - x \cdot P) - \kappa|b - x \cdot P|}{1 - \kappa^2} \\ &\leq b.\end{aligned}$$

So the claim is proved. Therefore the polar equation of the oval is then given by

$$h(x, P, b) = \rho_-(x) = \frac{(b - \kappa^2 x \cdot P) - \sqrt{\Delta(x \cdot P)}}{1 - \kappa^2}. \tag{12.20}$$

From the physical constraint for refraction we must have $x \cdot \left(\frac{P - h(x, P, b)x}{|P - h(x, P, b)x|} \right) \geq \kappa$, and from the equation of the oval we then get that to have refraction we need

$$x \cdot P \geq b. \tag{12.21}$$

Concluding with the case $\kappa < 1$, given $P \in \mathbb{R}^n$ and $\kappa|P| < b < |P|$, keeping in mind (12.23) and (12.21), a refracting oval is the set

$$\mathcal{O}(P, b) = \{h(x, P, b)\, x : x \in S^{n-1}, x \cdot P \geq b\}$$

where

$$h(x, P, b) = \frac{(b - \kappa^2 x \cdot P) - \sqrt{(b - \kappa^2 x \cdot P)^2 - (1 - \kappa^2)(b^2 - \kappa^2 |P|^2)}}{1 - \kappa^2}.$$

Remark 12.3 If $|P| \to \infty$, then the oval converges to an ellipsoid which is the surface having the uniform refraction property in the far field case, see Sect. 12.3.2.

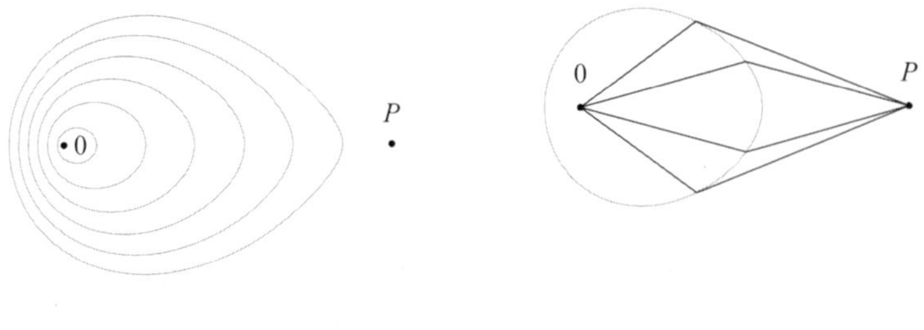

Cartesian ovals $\kappa < 1$, e.g., glass to air. (**a**) $|X| + 2/3|X - P| = 1.4 - 1.9$, $P = (2, 0)$. (**b**) $|X| + 2/3|X - P| = 1.7$, $P = (2, 0)$

In fact, if $m = P/|P|$ and $b = \kappa|P| + C$ with C a positive constant such that $\kappa|P| + C < |P|$, we can write

$$
\begin{aligned}
&h(x, P, b)\\
&= \frac{b^2 - \kappa^2|P|^2}{b - \kappa^2 x \cdot P + \sqrt{\Delta(x \cdot P)}}\\
&= \frac{C(2\kappa|P| + C)}{(\kappa|P| - \kappa^2 x \cdot m|P| + C) + \sqrt{(\kappa|P| - \kappa^2 x \cdot m|P| + C)^2 - (1 - \kappa^2)C(2\kappa|P| + C)}}\\
&\to \frac{2\kappa C}{(\kappa - \kappa^2 x \cdot m) + \sqrt{(\kappa - \kappa^2 x \cdot m)^2}} = \frac{C}{1 - \kappa x \cdot m}
\end{aligned}
$$

as $|P| \to \infty$.

12.6 Case $\kappa > 1$

In this case we must have $|P| \leq b$, and in case $b = |P|$ the oval reduces to the point P. Also $b < \kappa|P|$, since otherwise the points 0, P are inside the oval or 0 is on the oval, and therefore there cannot be refraction if $b \geq \kappa|P|$. So to have refraction we must have $|P| < b < \kappa|P|$ and so the point P is inside the oval and 0 is outside the oval.

Rewriting ρ in (12.16) we get that

$$
\rho_\pm(x) = \frac{(\kappa^2 x \cdot P - b) \pm \sqrt{(\kappa^2 x \cdot P - b)^2 - (\kappa^2 - 1)(\kappa^2|P|^2 - b^2)}}{\kappa^2 - 1}
$$

when $\Delta(x \cdot P) \geq 0$. Now $\Delta(t) \geq 0$ if and only if

$$t \notin \left(\frac{b - \sqrt{(\kappa^2 - 1)(\kappa^2 |P|^2 - b^2)}}{\kappa^2}, \frac{b + \sqrt{(\kappa^2 - 1)(\kappa^2 |P|^2 - b^2)}}{\kappa^2} \right),$$

and therefore $\rho_\pm(x)$ is well defined when $x \cdot P$ is not in this interval. If $x \cdot P \leq \dfrac{b - \sqrt{(\kappa^2 - 1)(\kappa^2 |P|^2 - b^2)}}{\kappa^2}$, then $\rho_\pm(x) \leq 0$. So to define $\rho_\pm(x)$ we must assume that $x \cdot P \geq \dfrac{b + \sqrt{(\kappa^2 - 1)(\kappa^2 |P|^2 - b^2)}}{\kappa^2}$. We have that $\rho_-(x) \leq \rho_+(x) \leq \dfrac{(\kappa^2 |P| - b) + \sqrt{\Delta(|P|)}}{\kappa^2 - 1} = \dfrac{\kappa |P| + b}{\kappa + 1} < b$. To have refraction, by the physical constraint we need to have $x \cdot \dfrac{P - \rho_\pm(x) x}{|P - \rho_\pm(x) x|} \geq 1/\kappa$, which is equivalent to $\kappa^2 x \cdot P - b \geq (\kappa^2 - 1)\rho_\pm(x)$. Therefore, the physical constraint is only satisfied by ρ_-.

Therefore if $\kappa > 1$, refraction only occurs when $|P| < b < \kappa |P|$, and the refracting piece of the oval is then given by

$$\mathcal{O}(P, b) = \left\{ h(x, P, b) x : x \cdot P \geq \frac{b + \sqrt{(\kappa^2 - 1)(\kappa^2 |P|^2 - b^2)}}{\kappa^2} \right\} \tag{12.22}$$

with

$$\begin{aligned} h(x, P, b) &= \rho_-(x) \\ &= \frac{(\kappa^2 x \cdot P - b) - \sqrt{(\kappa^2 x \cdot P - b)^2 - (\kappa^2 - 1)(\kappa^2 |P|^2 - b^2)}}{\kappa^2 - 1}. \end{aligned} \tag{12.23}$$

Remark 12.4 If $|P| \to \infty$, then the oval $\mathcal{O}(P, b)$ converges to the semi hyperboloid appearing in the far field refraction problem when $\kappa > 1$, see Sect. 12.3.3. Indeed, let $m = \dfrac{P}{|P|} \in S^{n-1}$ and $b = \kappa |P| - a$ with $a > 0$ a constant. Let

$$\Gamma(P, b) = \left\{ x \in S^{n-1} : x \cdot P \geq \frac{b + \sqrt{(\kappa^2 - 1)(\kappa^2 |P|^2 - b^2)}}{\kappa^2} \right\}.$$

For $x \in \Gamma(P, b)$ we have

$$\begin{aligned} &\frac{b + \sqrt{(\kappa^2 - 1)(\kappa^2 |P|^2 - b^2)}}{\kappa^2 |P|} \\ &= \frac{\kappa |P| - a + \sqrt{(\kappa^2 - 1)(\kappa^2 |P|^2 - (\kappa |P| - a)^2)}}{\kappa^2 |P|} \to \frac{1}{\kappa} \end{aligned}$$

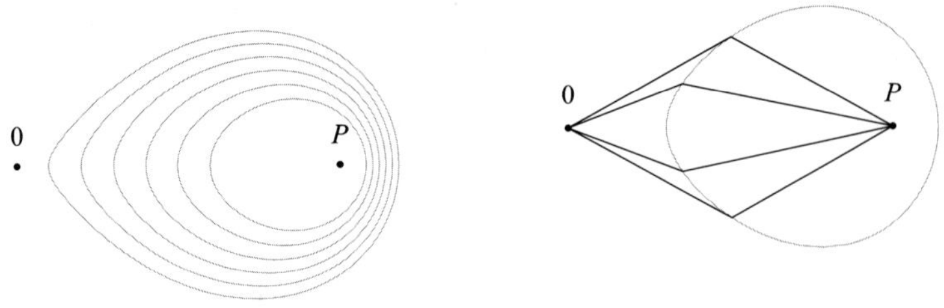

(a) (b)

Cartesian ovals $\kappa > 1$, e.g., air to glass. (**a**) $|X| + 3/2|X - P| = 2.9 - 2.4$, $P = (2, 0)$. (**b**) $|X| + 3/2|X - P| = 2.7$, $P = (2, 0)$

as $|P| \to \infty$. On the other hand, if $x \cdot m > 1/\kappa$, we get

$$
\begin{aligned}
&h(x, P, b)\\
&= \frac{a(2\kappa|P| - a)}{(\kappa^2|P|x \cdot m - \kappa|P| + a) + \sqrt{(\kappa^2|P|x \cdot m - \kappa|P| + a)^2 - (\kappa^2 - 1)a(2\kappa|P| - a)}}\\
&\to \frac{a2\kappa}{\kappa^2 x \cdot m - \kappa + \sqrt{(\kappa^2 x \cdot m - \kappa)^2}} = \frac{a}{\kappa x \cdot m - 1},
\end{aligned}
$$

as $|P| \to \infty$.

Chapter 13
Solution of the Far Field Refractor Problem $\kappa < 1$

Abstract This chapter solves the refractor problem in the far field. The notions of refractor measure, refractor mapping and weak solution are introduced. The refractor problem is then converted into an optimal transport problem and solved using the methods from Chap. 6. The chapter concludes with a brief description of recent related results.

Let n_1 and n_2 be the indexes of refraction of two homogeneous and isotropic media I and II, respectively. Suppose that from a point O inside medium I light emanates with intensity $f(x)$ for $x \in \Omega$. We want to construct a refracting surface $\mathcal{R}$ parameterized as $\mathcal{R} = \{\rho(x)x \, : \, x \in \overline{\Omega}\}$, separating media I and II, and such that all rays refracted by $\mathcal{R}$ into medium II have directions in Ω^* and the prescribed illumination intensity received in the direction $m \in \Omega^*$ is $f^*(m)$.

We first introduce the notions of refractor mapping and measure, and weak solution. We then convert the refractor problem into an optimal mass transport problem from $\overline{\Omega}$ to $\overline{\Omega^*}$ with the cost function $\log \dfrac{1}{1 - \kappa\, x \cdot m}$ and apply the results obtained before to establish existence and uniqueness of weak solutions.

Let Ω, Ω^* be two domains on S^{n-1}, the illumination intensity of the emitting beam is given by nonnegative $f(x) \in L^1(\overline{\Omega})$, and the prescribed illumination intensity of the refracted beam is given by a nonnegative Radon measure μ on $\overline{\Omega^*}$. Throughout this section, we assume that $|\partial\Omega| = 0$ and the physical constraint

$$\inf_{x\in\Omega, m\in\Omega^*} x \cdot m \geq \kappa. \tag{13.1}$$

We further suppose that the total energy conservation

$$\int_{\Omega} f(x)\, dx = \mu(\overline{\Omega^*}) > 0, \tag{13.2}$$

C. E. Gutiérrez, *Optimal Transport and Applications to Geometric Optics*, SpringerBriefs on PDEs and Data Science,
https://doi.org/10.1007/978-981-99-4867-3_13

and for any open set $G \subset \Omega$

$$\int_G f(x)\,dx > 0, \tag{13.3}$$

where dx denotes the surface measure on S^{n-1}.

13.1 Refractor Measure and Weak Solutions

We begin with the notions of refractor and supporting semi-ellipsoid.

Definition 13.1 A surface $\mathcal{R}$ parameterized by $\rho(x)x$ with $\rho \in C(\overline{\Omega})$ is a refractor from $\overline{\Omega}$ to $\overline{\Omega^*}$ for the case $\kappa < 1$ (often simply called refractor) if for any $x_0 \in \overline{\Omega}$ there exists a semi-ellipsoid $E(m, b)$ with $m \in \overline{\Omega^*}$ such that $\rho(x_0) = \dfrac{b}{1 - \kappa\, m \cdot x_0}$ and $\rho(x) \leq \dfrac{b}{1 - \kappa\, m \cdot x}$ for all $x \in \overline{\Omega}$. Such $E(m, b)$ is called a supporting semi-ellipsoid of $\mathcal{R}$ at the point $\rho(x_0)x_0$.

From the definition, any refractor is globally Lipschitz on $\overline{\Omega}$.

Definition 13.2 Given a refractor $\mathcal{R} = \{\rho(x)x : x \in \overline{\Omega}\}$, the refractor mapping of $\mathcal{R}$ is the multi-valued map defined by for $x_0 \in \overline{\Omega}$

$$\mathcal{N}_{\mathcal{R}}(x_0) = \{m \in \overline{\Omega^*} :\ E(m, b) \text{ supports } \mathcal{R} \text{ at } \rho(x_0)x_0 \text{ for some } b > 0\}.$$

Given $m_0 \in \overline{\Omega^*}$, the tracing mapping of $\mathcal{R}$ is defined by

$$\mathcal{T}_{\mathcal{R}}(m_0) = \mathcal{N}_{\mathcal{R}}^{-1}(m_0) = \{x \in \overline{\Omega} :\ m_0 \in \mathcal{N}_{\mathcal{R}}(x)\}.$$

Definition 13.3 Given a refractor $\mathcal{R} = \{\rho(x)x : x \in \overline{\Omega}\}$, the Legendre transform of $\mathcal{R}$ is defined by

$$\mathcal{R}^* = \{\rho^*(m)m :\ \rho^*(m) = \inf_{x \in \overline{\Omega}} \frac{1}{\rho(x)(1 - \kappa\, x \cdot m)},\quad m \in \overline{\Omega^*}\}.$$

We now give some basic properties of Legendre transforms.

Lemma 13.4 *Let $\mathcal{R}$ be a refractor from $\overline{\Omega}$ to $\overline{\Omega^*}$. Then*

(i) *$\mathcal{R}^*$ is a refractor from $\overline{\Omega^*}$ to $\overline{\Omega}$.*
(ii) *$\mathcal{R}^{**} = (\mathcal{R}^*)^* = \mathcal{R}$.*
(iii) *If $x_0 \in \overline{\Omega}$ and $m_0 \in \overline{\Omega^*}$, then $x_0 \in \mathcal{N}_{\mathcal{R}^*}(m_0)$ iff $m_0 \in \mathcal{N}_{\mathcal{R}}(x_0)$.*

Proof Given $m_0 \in \overline{\Omega^*}$, $\rho(x)(1 - \kappa x \cdot m_0)$ must attain the maximum over $\overline{\Omega}$ at some $x_0 \in \overline{\Omega}$. Then $\rho^*(m_0) = 1/[\rho(x_0)(1 - \kappa x_0 \cdot m_0)]$. We always have

$$\rho^*(m) = \inf_{x \in \overline{\Omega}} \frac{1}{\rho(x)(1 - \kappa m \cdot x)} \leq \frac{1}{\rho(x_0)(1 - \kappa x_0 \cdot m)}, \qquad \forall m \in \overline{\Omega^*}. \tag{13.4}$$

Hence $E(x_0, 1/\rho(x_0))$ is a supporting semi-ellipsoid to $\mathcal{R}^*$ at $\rho^*(m_0)m_0$. Thus, (i) is proved.

To prove (ii), from the definitions of Legendre transform and refractor mapping we have

$$\rho(x_0)\,\rho^*(m_0) = \frac{1}{1 - \kappa m_0 \cdot x_0} \quad \text{for } m_0 \in \mathcal{N}_{\mathcal{R}}(x_0). \tag{13.5}$$

For $x_0 \in \overline{\Omega}$, there exists $m_0 \in \mathcal{N}_{\mathcal{R}}(x_0)$ and so from (13.5) $\rho^*(m_0) = \dfrac{1/\rho(x_0)}{1 - \kappa x_0 \cdot m_0}$. By (13.4), $\rho^*(m)(1 - k x_0 \cdot m)$ attains the maximum $1/\rho(x_0)$ at m_0. Thus,

$$\rho^{**}(x_0) = \inf_{m \in \overline{\Omega^*}} \frac{1}{\rho^*(m)(1 - k x_0 \cdot m)} = \frac{1}{\rho(x_0)^{-1}}.$$

To prove (iii), we get from the proof of (ii) that if $m_0 \in \mathcal{N}_{\mathcal{R}}(x_0)$, then the semi-ellipsoid $E(x_0, 1/\rho(x_0))$ supports $\mathcal{R}^*$ at $\rho^*(m_0)m_0$ and so $x_0 \in \mathcal{N}_{\mathcal{R}^*}(m_0)$. On the other hand, if $x_0 \in \mathcal{N}_{\mathcal{R}^*}(m_0)$, we get that $m_0 \in \mathcal{N}_{\mathcal{R}^{**}}(x_0)$, and since $\mathcal{R}^{**} = \mathcal{R}$, $m_0 \in \mathcal{N}_{\mathcal{R}}(x_0)$. □

The next two lemmas discuss the refractor measure.

Lemma 13.5 *If $\mathcal{R}$ is a refractor, then $C = \{F \subset \overline{\Omega^*} : \mathcal{T}_{\mathcal{R}}(F)$ is Lebesgue measurable$\}$ is a σ-algebra containing all Borel sets in $\overline{\Omega^*}$.*

Proof Obviously, $\mathcal{T}_{\mathcal{R}}(\emptyset) = \emptyset$ and $\mathcal{T}_{\mathcal{R}}(\overline{\Omega^*}) = \overline{\Omega}$. Since $\mathcal{T}_{\mathcal{R}}(\cup_{i=1}^{\infty} F_i) = \cup_{i=1}^{\infty} \mathcal{T}_{\mathcal{R}}(F_i)$, C is closed under countable unions. Clearly for $F \subset \overline{\Omega^*}$ (since $\mathcal{N}_{\mathcal{R}}(x) \neq \emptyset$ for all $x \in \bar{\Omega}$)

$$\begin{aligned}
&\mathcal{T}_{\mathcal{R}}(F^c) \\
&= \{x \in \overline{\Omega} : \mathcal{N}_{\mathcal{R}}(x) \cap F^c \neq \emptyset\} \\
&= \{x \in \overline{\Omega} : \mathcal{N}_{\mathcal{R}}(x) \cap F^c \neq \emptyset,\ \mathcal{N}_{\mathcal{R}}(x) \cap F = \emptyset\} \\
&\quad \times \cup \{x \in \overline{\Omega} : \mathcal{N}_{\mathcal{R}}(x) \cap F^c \neq \emptyset,\ \mathcal{N}_{\mathcal{R}}(x) \cap F \neq \emptyset\} \\
&= [\mathcal{T}_{\mathcal{R}}(F)]^c \cup [\mathcal{T}_{\mathcal{R}}(F^c) \cap \mathcal{T}_{\mathcal{R}}(F)].
\end{aligned} \tag{13.6}$$

We show that the second term in this union has measure zero. If $x \in \mathcal{T}_{\mathcal{R}}(F^c) \cap \mathcal{T}_{\mathcal{R}}(F) \cap \bar{\Omega}$, then $\mathcal{R}$ parameterized by ρ has two distinct supporting semi-ellipsoids $E(m_1, b_1)$ and $E(m_2, b_2)$ at $\rho(x)x$. Notice that the set $\mathcal{T}_{\mathcal{R}}(F^c) \cap \mathcal{T}_{\mathcal{R}}(F) \cap \partial\Omega$ is a set of measure zero because of the assumption $|\partial\Omega| = 0$. We show that $\rho(x)x$ is a singular point of $\mathcal{R}$ for each $x \in \mathcal{T}_{\mathcal{R}}(F^c) \cap \mathcal{T}_{\mathcal{R}}(F) \cap \Omega$. Otherwise, if $\mathcal{R}$ has tangent hyperplane Π at $\rho(x)x$, then Π must coincide both with the tangent hyperplane of $E(m_1, b_1)$ and that of $E(m_2, b_2)$ at $\rho(x)x$. It follows from the Snell law that $m_1 = m_2$. Since ρ is Lipschitz, is differentiable a.e., and therefore, the area measure of $\mathcal{T}_{\mathcal{R}}(F^c) \cap \mathcal{T}_{\mathcal{R}}(F)$ is 0. So C is closed under complements, and we have proved that C is a σ-algebra.

To prove that C contains all Borel subsets, it suffices to show that $\mathcal{T}_{\mathcal{R}}(K)$ is compact if $K \subset \overline{\Omega^*}$ is compact. Let $x_i \in \mathcal{T}_{\mathcal{R}}(K)$ for $i \geq 1$. There exists $m_i \in \mathcal{N}_{\mathcal{R}}(x_i) \cap K$. Let $E(m_i, b_i)$ be the supporting semi-ellipsoid to $\mathcal{R}$ at $\rho(x_i)x_i$. We have

$$\rho(x)(1 - \kappa m_i \cdot x) \leq b_i \qquad \text{for } x \in \overline{\Omega}, \tag{13.7}$$

where the equality in (13.7) occurs at $x = x_i$. Since ρ is continuous in $\bar{\Omega}$ and distance from the source to the refractor is positive, we have that $0 < a_1 := \min_{\bar{\Omega}} \rho \leq \rho(x) \leq a_2 := \max_{\bar{\Omega}} \rho$ on $\overline{\Omega}$. By (13.7) and (13.1), $a_1(1 - \kappa) \leq b_i \leq a_2(1-\kappa^2)$. Then b_i contains a convergent subsequence. Since $\bar{\Omega}$ and K are compact, we may assume through subsequence that $x_i \longrightarrow x_0$, $m_i \longrightarrow m_0 \in K$, $b_i \longrightarrow b_0$, as $i \longrightarrow \infty$. By taking limit in (13.7), one obtains that the semi-ellipsoid $E(m_0, b_0)$ supports $\mathcal{R}$ at $\rho(x_0)x_0$ and $x_0 \in \mathcal{T}_{\mathcal{R}}(m_0)$. This proves $\mathcal{T}_{\mathcal{R}}(K)$ is compact. □

Lemma 13.6 *Given a refractor $\mathcal{R}$ and a nonnegative $f \in L^1(\overline{\Omega})$, the set function*

$$\mathcal{M}_{\mathcal{R},f}(F) = \int_{\mathcal{T}_{\mathcal{R}}(F)} f(x)\, dx$$

is a finite Borel measure defined on C and is called the refractor measure associated with $\mathcal{R}$ and f. Notice that $\mathcal{M}_{\mathcal{R},f}$ denotes the push forward measure (Definition 5.1) of $f\, dx$ through the refractor map $\mathcal{N}_{\mathcal{R}}$ given by Definition 13.2.

Proof Let $\{F_i\}_{i=1}^{\infty}$ be a sequence of pairwise disjoint sets in C. Let $H_1 = \mathcal{T}_{\mathcal{R}}(F_1)$, and $H_k = \mathcal{T}_{\mathcal{R}}(F_k) \setminus \cup_{i=1}^{k-1} \mathcal{T}_{\mathcal{R}}(F_i)$, for $k \geq 2$. Since $H_i \cap H_j = \emptyset$ for $i \neq j$ and $\cup_{k=1}^{\infty} H_k = \cup_{k=1}^{\infty} \mathcal{T}_{\mathcal{R}}(F_k)$, it follows that

$$\mathcal{M}_{\mathcal{R},f}(\cup_{k=1}^{\infty} F_k) = \int_{\cup_{k=1}^{\infty} H_k} f\, dx = \sum_{k=1}^{\infty} \int_{H_k} f\, dx.$$

Observe that $\mathcal{T}_{\mathcal{R}}(F_k) \setminus H_k = \mathcal{T}_{\mathcal{R}}(F_k) \cap (\cup_{i=1}^{k-1} \mathcal{T}_{\mathcal{R}}(F_i))$ is a subset of the singular set of $\mathcal{R}$ and has area measure 0 for $k \geq 2$. Therefore, $\int_{H_k} f\, dx = \mathcal{M}_{\mathcal{R},f}(F_k)$ and the σ-additivity of $\mathcal{M}_{\mathcal{R},f}$ follows. □

The notion of weak solutions is introduced through conservation of energy.

Definition 13.7 A refractor $\mathcal{R}$ is a weak solution of the refractor problem for the case $\kappa < 1$ with emitting illumination intensity $f(x)$ on $\overline{\Omega}$ and prescribed refracted illumination intensity μ on $\overline{\Omega^*}$ if for any Borel set $F \subset \overline{\Omega^*}$

$$\mathcal{M}_{\mathcal{R},f}(F) = \int_{\mathcal{T}_{\mathcal{R}}(F)} f\,dx = \mu(F). \tag{13.8}$$

13.2 Existence and Uniqueness of Solutions to the Refractor Problem

We introduce the cost

$$c(x,m) = \log \frac{1}{1 - \kappa\, x \cdot m}$$

for $x \in \Omega$ and $m \in \Omega^*$ where we assume $\Omega \cdot \Omega^* \geq \kappa$. From Definitions 6.1 and 13.1, $\mathcal{R} = \{\rho(x)x : x \in \overline{\Omega}\}$ is a refractor iff $\log \rho$ is c-concave. Using Definitions 6.3 and 13.2 we get that

$$\mathcal{N}_{c,\phi}(x) = \mathcal{N}_{\mathcal{R}}(x), \quad \text{with } \rho(z) = e^{\phi(z)}.$$

To apply the results from Chap. 6 and Sect. 6.1, we need to verify that the cost $c(x,m)$ satisfies (6.3). But from the Snell law and the proof of Lemma 13.5 this holds. From the definitions, $\mathcal{R}$ is a weak solution of the refractor problem iff $\log \rho$ is c-concave and $\mathcal{N}_{c,\log \rho}$ is a measure preserving mapping from $f(x)dx$ to μ in the sense of Definition 5.3. Hence from the energy conservation condition (13.2), we can apply Lemma 6.5, and so there exists a c-concave $\phi(x)$ such that (ϕ, ϕ^c) maximizes

$$I(u,v) = \int_{\overline{\Omega}} u f\,dx + \int_{\overline{\Omega^*}} v\,d\mu(m)$$

in $\mathcal{K} = \{(u,v) \in C(\overline{\Omega}) \times C(\overline{\Omega^*}) : u(x) + v(m) \leq c(x,m),\ \text{for } x \in \overline{\Omega}, m \in \overline{\Omega^*}\}$. Then from Lemma 6.4(ii), $\mathcal{N}_{c,\phi}(x)$ is a measure preserving mapping from $f\,dx$ to μ. Therefore, $\mathcal{R} = \{e^{\phi(x)}x : x \in \overline{\Omega}\}$ is a weak solution of the refractor problem completing the existence proof.

It remains to prove the uniqueness of solutions up to dilations. Let $\mathcal{R}_i = \{\rho_i(x)x : x \in \overline{\Omega}\}$, $i = 1, 2$, be two weak solutions of the refractor problem. It follows from Lemmas 6.4, 6.6 and 6.7 that $\mathcal{N}_{c,\log \rho_1}(x) = \mathcal{N}_{c,\log \rho_2}(x)$ a.e. on Ω. That is, $\mathcal{N}_{\mathcal{R}_1}(x) = \mathcal{N}_{\mathcal{R}_2}(x)$ a.e. on Ω. From the Snell law $\nu_i(x) = \dfrac{x - \kappa \mathcal{N}_{\mathcal{R}_i}(x)}{|x - \kappa \mathcal{N}_{\mathcal{R}_i}(x)|}$ is the unit normal to $\mathcal{R}_i$ pointing towards medium II at $\rho_i(x)x$ where $\mathcal{R}_i$ is differentiable. So $\nu_1(x) = \nu_2(x)$ a.e. We shall prove from this that $\rho_1(x) = C\rho_2(x)$ for some

$C > 0$. In fact, suppose x is a differentiability point of both ρ_1 and ρ_2 and also a point where the normals $\nu_1(x) = \nu_2(x)$. Take an orthogonal frame on the sphere S^{n-1}, for example, we can write points in S^{n-1} in spherical coordinates $x = x(\varphi_1, \cdots, \varphi_{n-1})$. At each differentiability point of ρ we will write the normal in terms of this frame. Since $|x(\varphi_1, \cdots, \varphi_{n-1})|^2 = 1$, we get that $x \cdot \partial_{\varphi_k} x = 0$ for $k = 1, \cdots, n-1$. Also by direct calculation $\partial_{\varphi_k} x \cdot \partial_{\varphi_j} x = 0$ when $k \neq j$. So the vectors $x, \partial_{\varphi_k} x, \cdots, \partial_{\varphi_{n-1}} x$ form a local orthogonal frame in S^{n-1} at each point. Let $X(\varphi_1, \cdots, \varphi_{n-1}) = x(\varphi_1, \cdots, \varphi_{n-1})\rho(x(\varphi_1, \cdots, \varphi_{n-1}))$. Each vector $\partial_{\varphi_k} X$, $1 \leq k \leq n-1$, is tangential to the surface parametrically defined by $X(\varphi_1, \cdots, \varphi_{n-1})$. Therefore $(\partial_{\varphi_k} X) \cdot \nu(x) = 0$, where $\nu(x)$ is the normal to the parametric surface $X(\varphi_1, \cdots, \varphi_{n-1})$. On the other hand,

$$\partial_{\varphi_k} X = (\partial_{\varphi_k} x)\, \rho(x) + \left(\sum_{j=1}^{n} \frac{\partial \rho}{\partial x_j}\, \partial_{\varphi_k} x_j \right) x,$$

$1 \leq k \leq n-1$. We can write the normal in terms of the frame as

$$\nu(x) = \sum_{\ell=1}^{n-1} a_\ell\, \partial_{\varphi_\ell} x + a_n\, x. \tag{13.9}$$

Therefore for $1 \leq k \leq n-1$

$$\begin{aligned}
0 = (\partial_{\varphi_k} X) \cdot \nu(x) &= \sum_{\ell=1}^{n-1} a_\ell\, (\partial_{\varphi_k} X) \cdot (\partial_{\varphi_\ell} x) + a_n\, (\partial_{\varphi_k} X) \cdot x \\
&= \sum_{\ell=1}^{n-1} a_\ell \left((\partial_{\varphi_k} x)\, \rho(x) + \left(\sum_{j=1}^{n} \frac{\partial \rho}{\partial x_j}\, \partial_{\varphi_k} x_j \right) x \right) \cdot (\partial_{\varphi_\ell} x) \\
&\quad + a_n \left((\partial_{\varphi_k} x)\, \rho(x) + \left(\sum_{j=1}^{n} \frac{\partial \rho}{\partial x_j}\, \partial_{\varphi_k} x_j \right) x \right) \cdot x \\
&= \rho(x) \sum_{\ell=1}^{n-1} a_\ell\, (\partial_{\varphi_k} x) \cdot (\partial_{\varphi_\ell} x) + \sum_{\ell=1}^{n-1} \left(\sum_{j=1}^{n} \frac{\partial \rho}{\partial x_j}\, \partial_{\varphi_k} x_j \right) x \cdot (\partial_{\varphi_\ell} x) \\
&\quad + a_n\, \rho(x)\, (\partial_{\varphi_k} x) \cdot x + a_n \left(\sum_{j=1}^{n} \frac{\partial \rho}{\partial x_j}\, \partial_{\varphi_k} x_j \right) x \cdot x \\
&= \rho(x)\, a_k\, |\partial_{\varphi_k} x|^2 + a_n \sum_{j=1}^{n} \frac{\partial \rho}{\partial x_j}\, \partial_{\varphi_k} x_j .
\end{aligned}$$

From (13.9), we get $a_n = x \cdot \nu(x)$, and therefore

$$a_k = -\frac{(x \cdot \nu(x)) \sum_{j=1}^{n} \dfrac{\partial \rho}{\partial x_j} \partial_{\varphi_k} x_j}{\rho(x)\,|\partial_{\varphi_k} x|^2} = -\frac{(x \cdot \nu(x))\, \partial_{\varphi_k} \log \rho\,(x(\varphi_1, \cdots, \varphi_{n-1}))}{|\partial_{\varphi_k} x|^2},$$

for $1 \leq k \leq n-1$. Now if $\nu_1(x) = \nu_2(x)$ a.e., then

$$\frac{(x \cdot \nu_1(x))\, \partial_{\varphi_k} \log \rho_1\,(x(\varphi_1, \cdots, \varphi_{n-1}))}{|\partial_{\varphi_k} x|^2} = \frac{(x \cdot \nu_2(x))\, \partial_{\varphi_k} \log \rho_2\,(x(\varphi_1, \cdots, \varphi_{n-1}))}{|\partial_{\varphi_k} x|^2}$$

a.e., implying

$$\partial_{\varphi_k} \log \rho_1\,(x(\varphi_1, \cdots, \varphi_{n-1})) = \partial_{\varphi_k} \log \rho_2\,(x(\varphi_1, \cdots, \varphi_{n-1})),$$

a.e. for $1 \leq k \leq n-1$, and we get $\log \rho_1\,(x(\varphi_1, \cdots, \varphi_{n-1})) = \log \rho_2\,(x(\varphi_1, \cdots, \varphi_{n-1})) + A$, for some constant A and for all x. We then obtain $\rho_1 = C\,\rho_2$ in Ω.

Summarizing, we have proved the following theorem.

Theorem 13.8 *Let Ω, Ω^* be subsets of S^{n-1} satisfying $\Omega \cdot \Omega^* \geq \kappa$ with $\kappa = n_2/n_1 < 1$; $|\partial\Omega| = 0$. Suppose we are given $f \in L^1(\Omega)$ with $f > 0$ a.e. and μ a Radon measure in Ω^* satisfying the conservation condition*

$$\int_\Omega f(x)\,dx = \mu(\Omega^*).$$

Then there exists refractor $\mathcal{R} = \{\rho(x)x : x \in \bar{\Omega}\}$, unique up to dilations, solution to $\mathcal{M}_{\mathcal{R},f} = \mu$.

13.3 Further Results

Concerning smoothness properties of refractors it can be shown that they gradients that are Hölder continuous of order α, for some $0 < \alpha < 1$, assuming appropriate conditions on the target and independently of the smoothness of the densities, see [31]. Earlier regularity results for reflectors, i.e., showing that the gradient is C^1, can be found in [10]. In Optics, when a ray is refracted, there is also a ray and part of the energy that is reflected back. These type of problems have been modeled and solved in [25], [27], and numerical schemes are introduced in [36]; further related results can be found in [29] and [28]. The Snell law has been recently extended in a generalized sense to deal with nano structures called metasurfaces. That is, on the surface Γ separating media I and II nano materials are arranged so that incident rays are refracted in unusual ways. This is modeled by introducing a function ϕ defined in a neighborhood of Γ, called a phase discontinuity, and the generalized Snell law has the form (12.2) but with the right hand side having the additive term $\nabla\phi$. For a proof of this generalization of the Snell law and applications see [26]. See also [30] for chromatic aberration questions, [4] for the design of reflecting metasurfaces with graphene, and [23] for the application of optimal transport to solve problems with metasurfaces.

Chapter 14
Proof of the Disintegration Theorem

Abstract This chapter contains a detailed proof of the disintegration theorem, Theorem 4.3, used in Chap. 4 to analyze the Wasserstein metric.

In this section, we prove the disintegration theorem, Theorem 4.3, that was used in Sect. 4. For completeness, we begin recalling the following results.

Lemma 14.1 *Let D be a linear dense subspace of a normed space X, Y is a Banach space, and T: D $\to$ Y is a bounded linear operator. Then T has a unique extension to a bounded linear operator $\hat{T} : X \to Y$ with $||\hat{T}|| = ||T||$.*

Lemma 14.2 *Let μ be a Borel measure on X, $T : X \to Y$ a Borel map, $\nu = T_{\#}\mu$, $f \in L^1(Y, \nu)$. Then*

$$\int_Y f(y)d\nu(y) = \int_X f(T(x))d\mu(x).$$

Lemma 14.3 *Let $(X, \mathcal{F})$ be a measurable space. Suppose $f_n : X \to \mathbb{R}$ is a sequence of $\mathcal{F}$-measurable maps that converges point-wise to f. Then f is a $\mathcal{F}$-measurable.*

Theorem 14.4 *Riesz representation theorem, [14, Section 6, p. 261]. Let X be a compact Hausdorff space. If $\psi : C(X) \to \mathbb{C}$ is a positive linear bounded functional, then there exits a unique regular complex Borel measure μ such that $\psi(f) = \int_X f(x)\,d\mu(x)$ for all $f \in C(X)$. As usual, $C(X)$ denotes the space of continuous functions in X with the maximum norm.*

Definition 14.5 We say that the normed spaces X and Y are isomorphic if there exists a linear bijection $T : X \to Y$ such that T and its inverse are continuous. Moreover, we say that X and Y are isometric if $||T(x)|| = ||x||$ for every $x \in X$.

We now prove the principal disintegration theorem, all others will be consequences of this one.

C. E. Gutiérrez, *Optimal Transport and Applications to Geometric Optics*, SpringerBriefs on PDEs and Data Science,
https://doi.org/10.1007/978-981-99-4867-3_14

Theorem 14.6 (Disintegration Theorem for Product of Compact Metric Spaces) *Let X and Y be compact metric spaces, and let μ be a finite Borel measure on $X \times Y$. Let $\pi_1 : X \times Y \to X$ be the projection onto the first coordinate, and set $\nu = (\pi_1)_{\#}\mu$.*

Then, there exists a unique a.e.-ν family of finite Borel measures $\{\mu_x\}_{x\in X}$ on Y such that

(a) For each Borel set E in Y, the map $x \mapsto \mu_x(E)$ is measurable.
(b) For any $g \in L^1(\mu)$,

$$\int_{X\times Y} g(x, y)d\mu(x, y) = \int_X \int_Y g(x, y)d\mu_x(y)d\nu(x). \tag{14.1}$$

Proof

Step 1 Given $\psi \in C(Y)$ and $\phi \in L^1(X, \nu)$ consider the bilinear form

$$B(\phi, \psi) = \int_{X\times Y} \phi(x)\psi(y)d\mu(x, y).$$

Then for each $\psi \in C(Y)$, the mapping $B(\cdot, \psi) : L^1(X, \nu) \to \mathbb{R}$ is a bounded linear operator, and by duality there exists a unique function $I(\psi) \in L^\infty(X, \nu)$ such that

$$B(\phi, \psi) = \int_X \phi(x)\, I(\psi)(x)\, d\nu(x) \tag{14.2}$$

for all $\phi \in L^1(X, \nu)$ and satisfying

$$\|B(\cdot, \psi)\| = \sup_{\|\phi\|_{L^1(X,\nu)}\leq 1} |B(\phi, \psi)| = \|I(\psi)\|_{L^\infty(X,\nu)}.$$

Since μ is a Borel measure in $X \times Y$, then ν is a Borel measure in X and so $I(\psi)$ is a Borel measurable function in X.

First, notice that $B(\phi, \psi)$ is well defined for $\psi \in C(Y)$ and $\phi \in L^1(X, \nu)$ because

$$|B(\phi, \psi)| \leq \int_{X\times Y} |\phi(x)||\psi(y)|d\mu(x, y) \leq ||\psi||_\infty \int_{X\times Y} |\phi(x)|d\mu(x, y)$$

$$= ||\psi||_\infty \int_X |\phi(x)|d\nu(x) = ||\psi||_\infty||\phi||_{L^1(X,\nu)} < \infty.$$

Hence $\|B(\cdot, \psi)\| \leq \|\psi\|_\infty$ and so $B(\cdot, \psi)$ is a bounded functional.

In addition, if $\phi_i \in C(Y)$ and $\lambda_i \in \mathbb{R}$, for $1 \leq i \leq N$, then

$$B\left(.,\sum_{i=1}^{N}\lambda_i\,\psi_i\right)=\sum_{i=1}^{N}\lambda_i\,B\left(\cdot,\psi_i\right)$$

implying that

$$I\left(\sum_{i=1}^{N}\lambda_i\,\psi_i\right)=\sum_{i=1}^{N}\lambda_i\,I(\psi_i),\,a.e. \tag{14.3}$$

Step 2 Since $C(Y)$ is separable, it has a countable dense subset $S = \{\psi_n\}$. Let span(S) be the linear space of all finite linear combinations of elements in S. Given $\psi \in S$ the function $I(\psi) \in L^\infty(X, \nu)$ so the evaluation $I(\psi)(x)$ is finite for a.e. $x \in X$ in the measure ν. We shall prove that there is a set N of ν-measure zero such that $I(\psi)(x)$ is well defined for all $x \in X \setminus N$ and for all $\psi \in S$. For each $x \in X \setminus N$, let $\nu_x : \text{span}(S) \to \mathbb{R}$ be defined by $\nu_x(\psi) = I(\psi)(x)$. *We show that for $x \in X \setminus N$, ν_x is linear and bounded in* span(S) *with respect to the norm of $C(Y)$ and therefore it extends as a linear mapping to the whole $C(Y)$.*

Let $X_{\psi_n} = \{x \in X : I(\psi_n)(x) \text{ is finite}\}$. Since $I(\psi_n) \in L^\infty(X, \nu)$, $\nu(X^c_{\psi_n}) = 0$. If $X_S = \bigcap_{n\in\mathbb{N}} X_{\psi_n}$, then $I(\psi_n)(x)$ is finite for all $x \in X_S$ and for all n, and $\nu(X^c_S) = 0$. Therefore, from (14.3) $I(\psi)(x)$ is finite for all $\psi \in \text{span}(S)$ and all $x \in X_S$, and ν_x is linear.

To show that ν_x is bounded, we have that $\|(\psi)\|_{L^\infty(X,\nu)} \leq \|\psi\|_\infty$ for all $\psi \in C(X)$ and so for $x \in X_S$

$$\begin{aligned}\|\nu_x\| &= \sup_{\|\psi\|_\infty\leq 1,\psi\in\text{span}(S)} |\nu_x(\psi)| = \sup_{\|\psi\|_\infty\leq 1,\psi\in\text{span}(S)} I(\psi)(x)| \\ &\leq \sup_{\|\psi\|_\infty\leq 1,\psi\in\text{span}(S)} \|\psi\|_\infty \leq 1.\end{aligned}$$

Since span(S) is dense in $C(Y)$, it follows from Lemma 14.1 that ν_x extends to a linear functional on $C(Y)$ for each $x \in X_S$; the extension denoted also by ν_x.

Step 3 *Since for every $x \in X_S$, ν_x is a bounded linear operator on $C(Y)$, we can apply Theorem 14.4 to get the measures μ_x which will be the ones we are looking for.*

Let $x \in X_S$. By Theorem 14.4 applied to ν_x, there is a unique Borel measure μ_x on Y such that

$$\nu_x(f) = \int_Y f(y)d\mu_x(y) \qquad \forall f \in C(Y).$$

For $x \in X \setminus X_S$, we set $\mu_x = 0$.

Note that for $x \in X_S$,

$$\mu_x(Y) = \int_Y d\mu_x(y) = \nu_x(1) = I(1)(x) \le \|I(1)\|_{L^\infty(X,\nu)} \le 1$$

and so μ_x is a sub-probability measure in Y, i.e. $\mu_x(E) \le 1$ for each Borel set $E \subset Y$.

Step 4 Given a Borel set $E \subset Y$, the function $\mu_E : X \to \mathbb{R}$ defined by $\mu_E(x) = \mu_x(E)$ is a Borel measurable function in X.

Given and open set $O \subset \mathbb{R}$,

$$\mu_E^{-1}(O) = \{x \in X : \mu_x(E) \in O\}.$$

Let $x \in X_S$. Since $C(Y)$ is dense in $L^1(Y, \mu_x)$, from Lemma 14.1 ν_x extends to $L^1(Y, \mu_x)$, with $\nu_x(f) = \lim_{n\to\infty} \nu_x(\psi_n)$ where ψ_n is a sequence in $C(Y)$ such that $\|\psi_n - f\|_{L^1(Y,\mu_x)} \to 0$ as $n \to \infty$. Now, since $1_E \in L^1(Y, \mu_x)$, there is a sequence $(\psi_n) \subset C(Y)$ such that $\|\psi_n - 1_E\|_{L^1(Y,\mu_x)} \to 0$. By Step 3,

$$\int_Y \psi_n(y) d\mu_x(y) = \nu_x(\psi_n) \qquad \forall n,$$

and so letting $n \to \infty$, we get

$$\mu_x(E) = \lim_{n\to\infty} \nu_x(\psi_n) = \lim_{n\to\infty} I(\psi_n)(x).$$

Let

$$L(x) = \begin{cases} \lim_{n\to\infty} I(\psi_n)(x) & x \in X_S \\ 0 & x \in X \setminus X_S \end{cases}$$
$$= \lim_{n\to\infty} \begin{cases} I(\psi_n)(x) & x \in X_S \\ 0 & x \in X \setminus X_S \end{cases}.$$

Since $I(\psi_n)$ are Borel measurable functions in X and X_S is a Borel set, by Lemma 14.3, $L(x)$ is Borel measurable. Thus,

$$\mu_E^{-1}(O) = \{x \in X : \mu_x(E) \in O\} = \{x \in X : L(x) \in O\} = L^{-1}(O)$$

is a Borel set. Therefore, part (a) of our theorem is proved. We proceed to show part (b).

(b) Note that for any $\psi \in S, \phi \in L^1(X, \nu)$, we have

$$B(\phi, \psi) = \int_{X\times Y} \phi(x)\psi(y)d\mu(x, y) = \int_X \phi(x)\, I(\psi)(x)d\nu(x) \quad \text{by (14.2)}$$

$$= \int_X \phi(x)\nu_x(\psi)d\nu(x) = \int_X \phi(x) \int_Y \psi(y)d\mu_x(y)d\nu(x) \quad \text{by Step 3}$$

$$= \int_X \int_Y \phi(x)\psi(y)d\mu_x(y)d\nu(x)$$

and so (14.1) holds for any $g(x, y) = \phi(x)\psi(y)$ with $\phi \in L^1(X, \nu)$, $\psi \in S$.

By the density of S in $C(Y)$, we get (14.1) for $g \in L^1(X, \nu) \times C(Y)$. Our goal is to show (14.1) for $g \in L^1(X \times Y, \mu)$. Since the set of simple functions in $L^1(\mu)$ is dense in $L^1(X \times Y, \mu)$, it is enough that (b) holds for box functions $1_A(x)1_B(y)$ where $A \subset X$ and $B \subset Y$ are Borel sets. This clearly holds because such box functions can be approximated by functions $g \in L^1(X, \nu) \times C(Y)$.

Finally, we show that the family of measures $(\mu_x)_{x\in X}$ is unique. Suppose there is another family of measures $(\mu'_x)_{x\in X}$ satisfying (a) and (b). Let $E \subset X$ and $F \subset Y$ be Borel sets. Then, by (b),

$$\int_{X\times Y} 1_E(x)1_F(y)d\mu(x, y) = \int_X \int_Y 1_E(x)1_F(y)d\mu_x(y)d\nu(x)$$

and

$$\int_{X\times Y} 1_E(x)1_F(y)d\mu(x, y) = \int_X \int_Y 1_E(x)1_F(y)d\mu'_x(y)d\nu(x).$$

Thus,

$$\mu(E \times F) = \int_E \mu_x(F)d\nu(x) = \int_E \mu'_x(F)d\nu(x),$$

for each Borel set $E \subset X$, and so $\mu_x(F) = \mu'_x(F)$ for ν- a.e. $x \in X$. □

Definition 14.7 A Polish space is a complete separable metric space.

Our next goal is to prove a disintegration theorem for product of two Polish spaces. To do so, we need the following definition and theorem.

Definition 14.8 A sequence (μ_n) of probability measures over X is said to be tight if for every $\epsilon > 0$, there exists a compact subset $K \subset X$ such that $\mu_n(X \setminus K) < \epsilon$ for every n.

Theorem 14.9 ([5, Theorem 8.6.2 (Prokhorov)]) *Suppose (μ_n) is a tight sequence of sub-probability measures over a Polish space X (i.e. $\mu_n(X) \leq 1$ for all n). Then, there exists a measure μ on X and a subsequence (μ_{n_k}) such that $\mu_{n_k} \rightharpoonup \mu$.*

Theorem 14.10 (Disintegration Theorem for Products of Polish Spaces) *Let X, Y be two Polish spaces. Let μ be a Borel finite measure on $X \times Y$. Let $\nu = (\pi_1)_\# \mu$, where $\pi_1 : X \times Y \to X$ is the projection onto the first coordinate. Then, there exists a unique family of finite Borel measures $\{\mu_x\}_{x \in X}$ on Y such that*

(a) *For any Borel set E in Y, the map $x \mapsto \mu_x(E)$ is measurable.*
(b) *For any $g \in L^1(\mu)$,*

$$\int_{X \times Y} g(x, y) d\mu(x, y) = \int_X \int_Y g(x, y) d\mu_x(y) d\nu(x).$$

Proof

(a) $X \times Y$ is a Polish space since X and Y are Polish spaces. Since μ is a regular Borel measure, for each $m \in \mathbb{N}$ there is a compact set $K_m \subset X \times Y$ such that $\mu(X \times Y \setminus K_m) < 1/m$. For each $m \in \mathbb{N}$, $K_m \subset \pi_1(K_m) \times \pi_2(K_m)$, and so

$$\mu\left(X \times Y \setminus \pi_1(K_m) \times \pi_2(K_m)\right) \leq \mu(X \times Y \setminus K_m) < 1/m$$

where $\pi_1 : X \times Y \to X$ and $\pi_2 : X \times Y \to Y$ are the projection maps. Let μ_m be the measure obtained restricting μ to the compact $\pi_1(K_m) \times \pi_2(K_m)$. We then apply Theorem 14.6 with $X \rightsquigarrow \pi_1(K_m)$, $Y \rightsquigarrow \pi_2(K_m)$, and $\mu \rightsquigarrow \mu_m$ to obtain a family $\{\mu_x^m\}_{x \in \pi_1(K_m)}$ of sub-probability measures on $\pi_2(K_m)$ satisfying the conclusion of that theorem. Next extend each measure μ_x^m to the whole of Y as zero in $Y \setminus \pi_2(K_m)$ and denote the resulting family of measures also with $\{\mu_x^m\}_{x \in \pi_1(K_m)}$. Also for $x \in X \setminus \pi_1(K_m)$ define $\mu_x^m = 0$. Then by definition $\{\mu_x^m\}_{x \in X}$ is tight, and so by Theorem 14.9, it has a weakly convergent subsequence $\mu_x^{m_k} \rightharpoonup \mu_x$, for some μ_x measure in Y. Now let $E \subset X \times Y$ be a Borel set. We have that the map $x \mapsto \mu_x^{m_k}(E)$ is a Borel map from $X \to \mathbb{R}$. Since $\mu_x^{m_k}(E) \to \mu_x(E)$ as $k \to \infty$ and the limit of Borel maps is a Borel map, we conclude that the map $x \mapsto \mu_x(E)$ is Borel measurable completing the proof of (a).

(b) Let $g \in L^1(X, \nu) \times C^0(Y)$, and write $g(x, y) = f(x)h(y)$, where $f \in L^1(X, \nu)$ and $h \in C^0(Y)$. By weak convergence, we know that

$$\int_Y h(y) d\mu_x^{m_k}(y) \to \int_Y h(y) d\mu_x(y).$$

Hence

$$\int_X f(x) \int_Y h(y) d\mu_x^{m_k}(y) d\nu(x) \to \int_X f(x) \int_Y h(y) d\mu_x(y) d\nu(x)$$

by Lebesgue Dominated Convergence Theorem which can be applied because $f \in L^1(X, \nu)$ and

$$\int_Y |h(y)| d\mu_x^{m_k}(y) = \int_K |h(y)| d\mu_x^{m_k}(y) \leq C\mu_x^{m_k}(K) \leq C$$

where K is the compact support of $h(y)$ and $C = ||h||_\infty$. On the other hand, let $g \in L^1(\mu)$ and $\epsilon > 0$. Then there exists $\delta > 0$ such that $\int_E |g(x, y)| d\mu(x, y) < \epsilon$ for $E \subset X \times Y$ with $\mu(E) < \delta$. Hence, if $m > 1/\delta$ we have $\mu(X \times Y \setminus K_m) < 1/m < \delta$ and so $\left| \int_{X \times Y \setminus K_m} g(x, y) d\mu(x, y) \right| < \epsilon$. Therefore

$$\int_{K_m} g(x, y) d\mu(x, y) \to \int_{X \times Y} g(x, y) d\mu(x, y),$$

as $m \to \infty$. But

$$\int_{K_{m_k}} g(x, y) d\mu(x, y) = \int_X \int_Y g(x, y) d\mu_x^{m_k}(y) d\nu(x)$$

for any $g \in L^1(\mu)$, and in particular for any $g \in L^1(X, \nu) \times C^0(Y)$. Therefore,

$$\int_{X \times Y} g(x, y) d\mu(x, y) = \int_X \int_Y g(x, y) d\mu_x(y) d\nu(x)$$

for any $g \in L^1(X, \nu) \times C^0(Y)$. From the approximation argument used in the proof of Theorem 14.6 we get our result for any $g \in L^1(\mu)$.

□

We end this section by showing the main disintegration theorem needed in the proof of Lemma 4.4 that was used to prove the triangle inequality for the Wasserstein distance.

Theorem 14.11 (Disintegration for Fibers of a Map) *Let X, Y be two Polish spaces. Let $f : Y \to X$ be a Borel map.*[1] *Let μ be a finite measure Borel measure on*

[1] In this formulation the roles of X and Y in Definition 4.2 are reversed.

Y, and $\nu = f_{\#}\mu$. Then, there exists a unique ν-a.e. family of finite Borel measures $\{\mu_x\}_{x\in X}$ on Y such that

(a) *For each Borel set E in Y, the map $x \mapsto \mu_x(E)$ is measurable.*
(b) *For ν-almost every $x \in X$, the measure μ_x is supported on the fiber $f^{-1}(x)$.*
(c) *For any $g \in L^1(\mu)$*

$$\int_Y g(y)d\mu(y) = \int_X \int_{f^{-1}(x)} g(y)d\mu_x(y)d\nu(x).$$

Proof

(a) Let $\gamma = (f \times id)_{\#}\mu$, i.e. for any Borel set E in $X \times Y$, $\gamma(E) = \mu((f \times id)^{-1}(E))$ where $f \times id : Y \to X \times Y$ is the map defined by $(f \times id)(y) = (f(y), y)$. Note that $\nu = (\pi_1)_{\#}\gamma$ because

$$\begin{aligned}(\pi_1)_{\#}\gamma(A) &= \gamma(\pi_1^{-1}(A)) = \gamma(A \times Y) = \mu((f \times id)^{-1}(A \times Y)) \\ &= \mu(f^{-1}(A)) = \nu(A)\end{aligned}$$

for any Borel set A in X. Theorem 14.10 applied to γ, one can find a family of finite measures $\{\mu_x\}_{x\in X}$ in Y such that for any Borel set E in Y, the map $x \mapsto \mu_x(E)$ is measurable. Moreover, for all $g \in L^1(\gamma)$,

$$\int_{X\times Y} g(x, y)d\gamma(x, y) = \int_X \int_Y g(x, y)d\mu_x(y)d\nu(x).$$

(b) Now, note that $\gamma\left(\{(f(y), y) : y \in Y\}^c\right) = 0$, and thus the last formula implies that supp $\mu_x \subset f^{-1}(x)$ for a.e. x.
(c) Finally, let $g \in L^1(\mu)$ and so $g \in L^1(\gamma)$ in the sense that $g(x, y) = g(y)$ for all $x \in X$. Then

$$\begin{aligned}\int_Y g(y)d\mu(y) = \int_{X\times Y} g(x, y)d\gamma(x, y) &= \int_X \int_Y g(x, y)d\mu_x(y)d\nu(x) \\ &= \int_X \int_{f^{-1}(x)} g(x, y)d\mu_x(y)d\nu(x)\end{aligned}$$

since supp $\mu_x \subset f^{-1}(x)$.

□

Reference

1. Benamou, J.-D., Brenier, Y.: A computational fluid mechanics solution to the Monge-Kantorovich mass transfer problem. Numerische Matematik **84**(3), 375–393 (2000)
2. Benamou, J.-D., Carlier, G., Nenna, L.: A numerical method to solve multi-marginal optimal transport problems with Coulomb cost. In: Glowinski, R., Osher, S.J., Yin, W. (eds.) Splitting Methods in Communication, Imaging, Science, and Engineering, pp. 577–601. Springer International Publishing, Cham (2016)
3. Beurling, A.: An automorphism of product measures. Ann. Math. **72**(1), 189–200 (1960)
4. Biswas, S.R., Gutiérrez, C.E., Nemilentsau, A., Lee, I.-H., Oh, S.-H., Avouris, P., Low, T.: Tunable graphene metasurface reflectarray for cloaking, illusion, and focusing. Phys. Rev. Appl. **9**, 034021 (2018)
5. Bolgachev, V.: Measure Theory, vol. 2. Springer, Berlin (2007)
6. Bourbaki, N.: Éléments de Mathématique, vol. Intégration, Chapitre 6, 2nd edn. Springer, Berlin (2007)
7. Brenier, Y.: Polar factorization and monotone rearrangement of vector-valued functions. Commun. Pure Appl. Math. **44**(4), 375–417 (1991)
8. Brualdi, R.: Combinatorial matrix classes. Encyclopedia of Mathematics and its Applications, vol. 108. Cambridge University Press, Cambridge (2006)
9. Caffarelli, L.A.: The regularity of mappings with a convex potential. J. Am. Math. Soc. **5**(1), 99–104 (1992)
10. Caffarelli, L.A., Gutiérrez, C.E., Huang, Q.: On the regularity of reflector antennas. Ann. Math. **167**, 299–323 (2008)
11. Chang J.T., Pollard, D.: Conditioning as disintegration. Stat. Neerlandica **51**(3), 287–317 (1997)
12. Dantzig, G.B.: Linear Programming and Extensions. Report R-366-PR, The Rand Corporation, Santa Monica, CA, 1963
13. Dellacherie, C., Meyer, P.-A.: Probabilities and Potential. North-Holland Mathematics Studies, vol. 29. North Holland, Amsterdam (1979)
14. Dunford, N., Schwartz, J.T.: Linear Operators, Part 1, General Theory, vol. 1. Wiley, New York (1958)
15. Evans, L.C., Gariepy, R.F.: Measure Theory and Fine Properties of Functions. CRC Press, Boca Raton (1992)
16. Essid, M., Pavon, M.: Traversing the Schrödinger bridge strait: Robert Fortet's marvelous proof redux. J. Optim. Theory Appl. **181**(1), 23–60 (2019)

C. E. Gutiérrez, *Optimal Transport and Applications to Geometric Optics*,
SpringerBriefs on PDEs and Data Science,
https://doi.org/10.1007/978-981-99-4867-3

17. Evans, L.C.: Partial differential equations and Monge–Kantorovitch mass transfer (2001). http://math.berkeley.edu/~evans/Monge-Kantorovich.survey.pdf
18. Fortet, R.: Résolution d'un système d'equations de M. Schrödinger. J. Math. Pure Appl. **IX**, 83–105 (1940)
19. Franklin, J.N.: Methods of Mathematical Economics. Classics in Applied Mathematics, vol. 37. SIAM, Philadelphia (2002)
20. Fulkerson, D.R.: Hitchcock transportation problem. Technical Report P-890, The Rand Corporation, Santa Monica, CA, 1956
21. Gutiérrez, C.E., Huang, Q.: The refractor problem in reshaping light beams. Arch. Ratio. Mech. Anal. **193**(2), 423–443 (2009)
22. Gutiérrez, C.E., Huang, Q.: The near field refractor. Annales de l'Institut Henri Poincaré (C) Analyse Non Linéaire **31**(4), 655–684 (2014)
23. Gutiérrez, C.E., Huang, Q., Mérigot, Q., Thibert, B.: Metasurfaces and optimal mass transport. SMAI J. Comput. Math. **8**, 201–224 (2022)
24. Ghys, É.: Gaspard Monge, report on cuttings and embankments (2012). https://images.math.cnrs.fr/Gaspard-Monge,1094.html?lang=fr.
25. Gutiérrez, C.E., Mawi, H.: The far field refractor with loss of energy. Nonlinear Anal. Theory Meth. Appl. **82**, 12–46 (2013)
26. Gutiérrez, C.E., Pallucchini, L., Stachura, E.: General refraction problems with phase discontinuities on nonflat metasurfaces. J. Opt. Soc. Am. A **34**(7), 1160–1172 (2017)
27. Gutiérrez, C.E., Sabra, A.: The reflector problem and the inverse square law. Nonlinear Anal. Theory Methods Appl. **96**, 109–133 (2014)
28. Gutiérrez, C.E., Sabra, A.: Aspherical lens design and imaging. SIAM J. Imaging Sci. **9**(1), 386–411 (2016). Preprint http://arxiv.org/pdf/1507.08237.pdf
29. Gutiérrez, C.E., Sabra, A.: Freeform lens design for scattering data with general radiant fields. Arch. Ration. Mech. Anal. **228**, 341–399 (2018)
30. Gutiérrez, C.E., Sabra, A.: Chromatic aberration in metalenses. Adv. Appl. Math. **124** (2021)
31. Gutiérrez. C.E., Tournier, F.: Regularity for the near field parallel refractor and reflector problems. Calc. Var. PDEs **54**(1), 917–949 (2015)
32. Gutiérrez, C.E.: The Monge–Ampère Equation. Progress in Nonlinear Differential Equations and Their Applications, vol. 89, 2nd edn. Birkhäuser, Boston (2016)
33. Gutiérrez, C.E., van Nguyen, T.: On Monge–Ampère type equations arising in optimal transportation problems. Calc. Var. Partial Differ. Equ. **28**(3), 275–316 (2007)
34. Huyghens, C.: Traité de la lùmiere, 1920 edn. Gauthier-Villars, Paris (1690). http://gallica.bnf.fr/ark:/12148/bpt6k5659616j/f1.image
35. Koopmans, T.C.: Optimum utilization of the transportation system. Econometrica (pre-1986) **17**(supplement), 136 (1949)
36. De Leo, R., Gutiérrez, C.E., Mawi, H.: On the numerical solution of the far field refractor problem. Nonlinear Anal. Theory Methods Appl. **157**, 123–145 (2017)
37. Ford, L.R. Jr., Fulkerson, D.R.: Flows in Networks. Princeton University Press, Princeton (1962)
38. Luneburg, R.K.: Mathematical Theory of Optics. University of California Press, Berkeley (1964)
39. Maxwell, J.C.: On the description of oval curves, and those having a plurality of foci; with remarks from Prof. Forbes;. Proc. R. Soc. Edin. **II**, 1–3 (1846). https://archive.org/details/scientificpapers01maxw
40. Maxwell, J.C.: In: Harman, P.M. (eds.) The Scientific Letters and Papers of James Clerk Maxwell, vol. 1, pp. 1846–1862. Cambridge University Press, Cambridge (1990)
41. Menon, M.V.: Reduction of a matrix with positive elements to a doubly stochastic matrix. Proc. Am. Math. Soc. **18**, 244–247 (1967)
42. Milnor, J.W.: Topology from the Differentiable Viewpoint. Princeton Landmarks in Mathematics, Princeton University Press, Princeton (1997)
43. Munkres, J.: Algorithms for the assignment and transportation problems. J. Soc. Indust. Appl. Math. **5**(1), 32–38 (1957)

44. Newton, S.I.: Newton's Principia, the Mathematical Principles of Natural Philosophy. First American Edition 1846 edn., Daniel Adee, New York (1687). https://archive.org/details/newtonspmathema00newtrich.
45. Orden, A.: The transhipment problem. Manag. Sci. **2**(3), 276–285 (1956)
46. Peyré, G., Cuturi, M.: Computational Optimal Transport: with Applications to Data Science. Foundations and Trends in Machine Learning Series, vol. 37. Now Publishers, Hanover (2019)
47. Rüschendorf, L.: Convergence of the iterative proportional fitting procedure. Ann. Stat. **23**(4), 1160–1174 (1995)
48. Rockafellar, R.T.: Convex Analysis. Princeton Landmarks in Mathematics and Physics. Princeton University Press, Princeton (1997)
49. Rachev, S.T., Rüschendorf, L.: Mass Transportation Problems, vols. I, II. Springer, New York (1998)
50. Santambrogio, F.: Optimal Transport for Applied Mathematicians. Progress in Nonlinear Differential Equations and Their Applications, vol. 87. Birkhäuser, Boston (2015)
51. Schrödinger, E.: Sur la théorie relativiste de l'électron et l'interprétation de la mécanique quantique. Ann. l'inst. Henri Poincaré **2**(4), 269–310 (1932)
52. Schneider, R.: Convex bodies: the Brunn–Minkowski theory. Encyclopedia of Applied and Computational Mathematics, vol. 44. Cambridge University Press, Cambridge (1993)
53. Sinkhorn, R.: A relationship between arbitrary positive matrices and doubly stochastic matrices. Ann. Math. Stat. **35**, 876–879 (1964)
54. Urbas, J.: Mass Transfer Problems. Reproduced as Ms. ed. Bonn: Collaborative Research Center 256 (1998)
55. C. Villani, *Topics in optimal transportation*, Graduate Studies in Mathematics, vol. 58, American Mathematical Society, Providence, RI, 2003.
56. von Neumann, J.: Zur operatorenmethode in der klassischen mechanik. Ann. Math. **33**(3), 587–642 (1932)

Printed in the United States
by Baker & Taylor Publisher Services